# African science and technology education into the new millennium: practice, policy and priorities

*Editors*

Prem Naidoo
Mike Savage

*A project publication by the African Forum for Children's Literacy in Science and Technology (AFCLIST)*

Juta

First published 1998



PO Box 14373, Kenwyn 7790



ISBN 0 7021 4476 2

Cover design: Abdul Amien, Cape Town
Sub-editing: John Linnegar
Book design and typesetting: Charlene Bate, Cape Town
Printed and bound in the Republic of South Africa by
The Rustica Press, Old Mill Road, Ndabeni, Western Cape
D6767

The African Forum for Children's Literacy in Science and Technology would like to dedicate this book in memory of Professor Rosalind Driver. She was a board member of AFCLIST who unselfishly gave her time to the development of quality science education in Africa and the world. Her contributions to science education, particularly on how children learn, are seminal and will continue to guide present and future research in the field of learning and science education.


## Acknowledgments

Many people have helped to make this book possible. We are particularly grateful to the discussants and Sidney Westley. Shakila Thakurpersad and Lucky Khumalo performed the hidden task of checking the references and tables.

Without the initiative and energy of AFCLIST and the generous support of the Rockefeller Foundation there would have been neither the African Science and Technology Education (ASTE '95) meeting nor this book. Other donors whose support made the meeting possible are the Norwegian Agency for Development (NORAD), the Foundation for Research Development (FRD), South Africa, and the International Development Research Council (IDRC). The University of Durban-Westville and its staff were exceptionally warm hosts whose contributions to the meeting must be fully acknowledged.

*Prem Naidoo*
*Mike Savage*
September 1998

# Preface

African educators and overseas friends came together from 4 to 9 December 1995 — about 100 from four continents and 14 countries, women and men, their ages from under 30 to over 80. Included were ministry officials and university administrators, scientists and classroom teachers, innovators or researchers into teaching, and teachers of teachers. Eleven main papers, authored in advance and by Africans, were the basis of our discussion, though all participants spoke as critics, proponents, and commentators. The lively discourse covered an amazing variety of concerns in the service of science and technology education. That topic addresses both the genetic system of that organism within society and the public subsoil that must nourish it.

No children took part (a few wandered by). Yet they are the main actors. Each evening we had a brief glimpse of today's practice in children's science. The African Forum for Children's Literacy in Science and Technology (AFCLIST), an activity of the Rockefeller Foundation, a major sponsor of the meeting, collaborated with our university host to show us what it is doing. The Forum is explicit on one issue: gender equity is a part of all the work it supports.

- Paper Making Educational Trust (PAMET), a project in Malawi, encourages primary schoolchildren to recycle paper to make products such as notebooks. They become involved in science and the technology of scaled-up production. This has become a significant income-generating project.
- In the Zanzibar Science Camps, cabinet ministers, scientists, education officers, teachers and children spend three weeks each year struggling with problems of science education. A major contribution one year was that of a young secondary schoolgirl when she exclaimed after a visit to a mangrove swamp, 'You know, we have to learn the language of trees.'
- 'Spider's Place' is a television series for younger children in South Africa. Spider, the leader of a gang of puppet children, is a girl. Their scientific and technological ingenuity gets the gang out of many a scrape.
- In Ghana a group of educators, scientists, teachers, students and industrialists became concerned at the lack of connection of school science with products such as aluminium cooking utensils, beer, charcoal and fertilizer that are found in every African village. Through a series of lively and intensive workshops they are producing an elegant collection of resource materials for science teachers and learners.

AFCLIST believes that involvement in the culture of science provides the youth with opportunities to participate actively in democratizing the educational process and society, and provides a base for the development of higher-level human resources in science and technology. We hope that the publication of this book advances the involvement in this culture of young people throughout the continent of Africa.

*Philip Morrison*
Emeritus Professor, Massachusetts Institute of Technology

## Biographical details of authors

### *Prof John D Volmink*

John D Volmink is currently director of the Centre for the Advancement of Science and Mathematics Education (CASME), which is based at the University of Natal, Durban. He is also acting Head of the University Education Development Programme.

He is a graduate of the University of Western Cape (UWC), where he completed his BSc and BSc (Hons). He later went to the USA, where he completed an MSc and a PhD in Mathematics Education. His research interests are in the cognitive and social aspects of mathematics education as well as assessment and evaluation.

Professor Volmink started his career as a high school teacher of science and mathematics. Thereafter he taught at the Peninsula Technikon, where he became Head of the Department of Mathematical Sciences. He later lectured in Applied Mathematics at UWC and the University of Cape Town.

Since the completion of his PhD studies he has also worked as assistant professor of Mathematics Education at Cornell University. He then returned to southern Africa and worked for a short while at the University of Botswana.

Since his return to South Africa, he has served on several national educational structures. During 1993 he was chairperson of the Southern African Association of Research in Mathematics and Science Education (SAARMSE). He is also deeply involved in community structures and in-service education.

### *Dr Marissa Rollnick*

Marissa Rollnick is a senior lecturer at the University of the Witwatersrand, where she is responsible for the chemistry section of the College of Science, an access programme for underprepared students. Prior to that, she worked in Swaziland for 15 years, first in a teacher-training college and then in the Education Faculty of the University of Swaziland. Her research interests are primarily in the area of cognition and language in Science Education.

### *Ms Vijay Reddy*

Vijay Reddy is a science educator. She has taught chemistry at high school, college of education and university. She has also worked in nongovernmental organizations (NGOs) involved in in-service education for science teachers, and in an evaluation and monitoring NGO. Her interests include issues of cognition in learning science and redress and equity in the field of research in South Africa. Her present research involves developing the life histories of South African black scientists.

### *Ms Karen Worth*

Karen Worth began her career as a teacher of young children in New York City and Boston and she continues to work closely with teachers and children in classrooms. She has extensive experience in elementary science education. She worked as curriculum and staff developer for both the Elementary Science Study (ESS) and the

African Primary Science Program (APSP) at the Education Development Centre in Africa in the 1960s. More recently, she was the principal investigator for the development of the Insights curriculum. She chaired the Working Group on Science Teaching Standards for the National Science Education Standards effort of the National Academy of Science Education and is currently co-director of the Centre for Urban Science Education Reform at the Education Development Centre, Inc, New York. She has also been a member of the faculty of the Wheelock College for over 25 years, where she teaches at the graduate school, and serves as consultant and adviser to the Boston Public Schools on staff and curriculum development at the elementary level and on science education reform. She is co-director of Wheelock's effort in pre-service collaboration in mathematics and science education funded by the National Science Foundation.

*Prof Emmanuel Fabiano*

Emmanuel Fabiano is the Deputy Director of AFCLIST. He is also the Principal of Chancellor College in Zomba, Malawi. He has been a secondary school teacher, a university science educator and a research chemist. Professor Fabiano has been a consultant for his government, UNESCO, UNDP, USAID and other organisations.

*Prof EA Yoloye*

EA Yoloye is an emeritus professor of Education of the University of Ibadan, Nigeria. For several years he taught chemistry at the CMG Grammar School in Lagos, Nigeria. He later took up an appointment as lecturer in Science Education at the Institute of Education, University of Ibadan, where he rose to the status of professor. At graduate level, he studied psychology, specializing in educational and psychological measurement and evaluation. He has had extensive experience in science education, curriculum development and evaluation. He coordinated the evaluation of the Primary Science Education Programme for Africa (SEPA) and he established the International Centre for Education Evaluation (ICEE) at the University of Ibadan. For 10 years he was the chairperson of the African Curriculum Organization (ACO). On retiring from active university teaching in 1989, he established the Amoye Institute for Educational Research and Development in Ibadan. He is currently chairperson of the Grants Committee and member of the Advisory Board of the African Forum for Children's Literacy in Science and Technology (AFCLIST).

*Prof Olugbemiro Jegede*

Olugbemiro Jegede is the head of the Research and Evaluation Unit, Distance Education Centre, University of Southern Queensland, Australia. He holds the degrees of BScEd and MEd from Ahmadu Bello University, Nigeria, and a PhD from the University of Wales, UK. Professor Jegede is also a chartered biologist of the London Institute of Biologists and a distinguished member of the New York Academy of Sciences. He was the foundation professor and dean of Education at the University of Abuja, Nigeria. Prior to this he was associate professor of Science

Education and held the positions of assistant dean, Faculty of Education, and head of Science Education at Ahmadu Bello University, where he worked for 17 years. His areas of interest include cultural studies, applied cognitive science, science education, computer-mediated communication, instructional design, distance education, research methodology, and sociocultural factors in non-Western environments. A recipient of the 1995 United States Quality Award for Excellence in Research and a 1996 Fellowship Award of the Science Teachers' Association of Nigeria for his contribution to science education globally, Prof Jegede has over 150 publications to his credit, including six books, chapter contributions to books, refereed journal articles, and refereed conference proceedings. Professor Jegede is a consultant for the UNDP (United Nations Development Program) and the Commonwealth Secretariat on Science, Technology and Environmental Education.

*Prof Gilbert Onwu*

Gilbert Onwu is a professor of Science Education and head of the Science and Maths Education Unit in the Department of Teacher Education at the University of Ibadan, Nigeria. With a background in chemistry and science education, he teaches courses in the departmental BEd, PGCE and higher degree (MEd, MPhil, PhD) programmes in science education. He received his BSc and PGCE from Goldsmiths College, University of London, and an MSc and a PhD in chemical education from the School of Chemical Sciences, University of East Anglia. His science-education research interests have focused on cognitive processes, with particular reference to problem-solving, learning difficulties in science, science process skills development/assessment and patterns of classroom transactions in large classes. Recently he has been interested in a cross-cultural dimension of these problems. Also, he has been working on innovative ways of teaching science to large classes using local scientific resources and a minimum of equipment. He has many publications to his credit, all of which have appeared in journals, books as well as monographs and technical reports. He has served as external examiner to a number of Nigerian universities and as consultant, resource person or expert to national education agencies, the Commonwealth Secretariat (CFTC), UNESCO, UNDP, WHO, etc. He is a member of the AFCLIST grants committee. He is currently on sabbatical leave, as a visiting professor in the Department of Mathematics and Science Education at the University of Venda.

*Mr Prem Naidoo*

Prem Naidoo, the director of AFCLIST, has been a secondary school teacher, a university lecturer, director of a university-based policy research unit, and is now the director of the Scholarship and Grant Funding of South Africa's Human Sciences Research Council (HSRC). An activist throughout his professional life, Prem believes that action must be informed and reflectively analysed, and that the process must involve all stakeholders. He has published a range of material and reports.

*Prof Mike Savage*

Mike Savage has taught at primary, secondary and tertiary levels. He has been a curriculum developer for projects in many African countries as well as in the United Kingdom and the United States of America. Savage has consulted for health, education and development projects supported by a wide range of donor organizations. He has edited many educational books, meeting proceedings and consultant reports.

*Dr Tom Mschindi*

Tom Mschindi, 37, is currently the managing editor of the *Daily Nation*, one of the publications published by the Nation Newspaper Ltd in Nairobi, Kenya. He has a keen interest in developmental journalism and finds time to read and contribute to scholarly journals on diverse topics in developmental journalism. He has published in the Fletcher Forum for World Affairs and in the Communication Training modules prepared by the African Council for Communication Education (ACCE).

He was educated in Nairobi University, from where he graduated Bachelor of Arts in Communication Studies, with distinction. He has attended several relevant courses and is busy setting up the Eastern Africa Media Institute, an International NGO to promote the development freedom and diversity of media in the East African region.

*Prof Hubert Dyasi*

Hubert Dyasi is professor of Science Education and director of the City College (City University of New York) where he also serves as director of the Workshop Center, a science-teacher development unit of the College. In addition to teaching undergraduate and graduate science education at the City College, Professor Dyasi conducts inquiry-based professional development programmes for teachers of selected schools and the community school district in New York City. He has wide international experience in science education, having served as the first executive director of the Science Education Program for Africa (SEPA) and as one of the developers of the United States National Science Education Standards and Assessments. He is a member of numerous advisory boards of American science education development programmes, and a science education consultant in South Africa.

# Abbreviations and acronyms

| | |
|---|---|
| ACO | African Curriculum Organization |
| AFCLIST | African Forum or Children's Literacy in Science and Technology |
| AMP | African Mathematics Programme |
| APSP | African Primary Science Programme |
| ASTE '95 | African Science and Technology Education, 1995 meeting |
| BOTSCI | Botswana Science |
| BSCS | Biological Sciences Curriculum Study |
| CASME | Centre for Advancement of Science and Mathematics Education |
| CASTME | Commonwealth Association of Science, Technology and Mathematics Educators |
| CBA | Chemical Bond Approach |
| CGIAR | Consultancy Group in International Agricultural Research |
| CIDA | Canadian International Development Agency |
| COPE | Community Orientated Primary Education |
| CUSO | Canadian University Service Overseas |
| DAAD | Deutscher Akademischer Austauschdienst |
| DANIDA | Danish International Development Agency |
| DSE | German Foundation for International Development |
| ECA | Economic Commission for Africa |
| EDC | Education Development Center (USA) |
| EEC | European Economic Community |
| Endicott House | African Education Programme Conference held in the USA in 1961, funded by USAID |
| ESS | Elementary Science Study |
| EU | European Union |
| FRD | Foundation for Research Development |
| GASAT 8 | Eighth International Gender and Science and Technology Conference |
| GER | Gross Enrolment Rate |
| GNP | Gross National Product |
| IBRD | International Bank of Reconstruction and Development |
| ICEE | International Centre for Educational Evaluation |
| ICIPE | International Centre for Insect Physiology and Entomology |
| IDA | International Development Agency |
| IDRC | International Development Research Council/Centre |
| IEA | International Education Association |
| ILO | International Labour Organization |
| IITA | International Institute for Tropical Agriculture |
| IMF | International Monetary Fund |
| IMSTIP | In-service Maths Science Improvement Programme |

| | |
|---|---|
| KCPE | Kenyan Certificate of Primary Education |
| KSTC | Kenya Science Teachers' College |
| KWPCS | Kagera Writers' and Publishers' Cooperative Society |
| MPSP | Mid-West State Primary Science Project |
| NGO | Nongovernmental organization |
| NORAD | Norwegian Agency for Development |
| NEPI | National Education Policy Initiative |
| NETF | National Education and Training Forum |
| NPE | National Policy on Education |
| NSF | National Science Foundation |
| NSSS | Nuffield Secondary School Science |
| ODM | Overseas Development Ministry |
| OECD | Organization for Economic Cooperation and Development |
| PAMET | Paper Making Educational Trust |
| PSSC | Physical Sciences Study Committee |
| SAARMSE | Southern African Association of Research in Mathematics and Science Education |
| SAP | Structural Adjustment Programme |
| SCIS | Science Education Improvement Study |
| SCISA | Science Curriculum Initiative in South Africa |
| SEP | Science Education Project |
| SEPA | (African Primary) Science Education Programme for Africa |
| SETC | Science Teacher Educators' Programme |
| SIDA | Swedish International Development Agency |
| SMSG | School Mathematics Study Group |
| STAG | Science and Technology in Action in Ghana |
| STAN | Science Teachers' Association of Nigeria |
| STS | Science and Technology in Society |
| TIMMS | Third International Measurement of Mathematics and Science |
| UNDP | United Nations Development Programme |
| UNECA | United Nations Economic Commission for Africa |
| UNEP | United Nations Environmental Program |
| UNESCO | United Nations Educational, Scientific and Cultural Organization |
| UNICEF | United Nations Children's Fund |
| UPE | Universal Primary Education |
| USAID | United States Agency for International Development |
| VSO | Voluntary Service Organization |
| ZIMSCI | Zimbabwe Science |

# Introduction

*Prem Naidoo and Mike Savage*

Worldwide, science and technology education has been advocated as an essential prerequisite for modernization and economic development (Forum; OECD, 1996). In Africa, countries recognized their importance and made them integral subjects in the curriculum from primary to tertiary education. Has science and technology education delivered on the claim of modernization and economic development? The impact has been disappointing. If anything, the people of Africa are suffering more than they were four decades ago. There is less inquiry science learning and more rote learning. Children are less rather than more able to extract meaning from their schooling in ways that can be applied to bring change to their lives. Thoughts that schooling could and should be enjoyable and linked to indigenous knowledge bases have become unthinkable.

The next millennium is upon us. Having made a disappointing impact in the past, can science and technology education meet the challenges of the coming century? Can we learn from legacies of the past to better shape the future?

The meeting organizers selected key areas of concern to help focus the analysis and provide guidelines for future practice, policy and priorities. This book reviews and analyses the legacies of science and technology education in sub-Saharan Africa.

**Chapter 1: Historical perspectives and their relevance to present and future practice, by EA Yoloye, Nigeria**

This chapter examines the historical perspectives of the last three decades and their relevance to the present and future of science and technology education. It pays particular attention to landmark meetings and organizations that had an impact on the continent. The chapter draws lessons from such organizations for the future, both at policy and at practice level.

**Chapter 2: The role of science and technology in development, by PM Makhurane, Zimbabwe, and M Kahn, South Africa**

The authors begin by presenting a historical perspective on the role of science and technology worldwide, with particular reference to Africa. They address questions such as: Is development linked to social and economic systems? Who defines development? For what kind of development should Africa strive? What kind of science and technology education best promotes this development? What is the relationship between science and technology and development? Do realistic or deterministic views of science and technology better suit development in Africa? The chapter provides evidence to support claims, analyses trends in the role of science and technology in development for past and current practices, and proposes suggestions for Africa in the future.

**Chapter 3: Curriculum innovations and their impact on the teaching of science and technology, by MBR Savage, Kenya**

This chapter examines curriculum innovations and their impact on the teaching of science and technology. It uses anecdotes to examine issues such as inquiry learning as a goal of curriculum change; curriculum change models; people development versus product development; holistic versus piecemeal innovation; teacher education models in relation to curriculum innovation and effective teaching; evaluation and assessment models; teaching in large classes and other constraining circumstances; the role of mass media models in change; and exemplars of AFCLIST-supported projects. The analysis of this chapter is framed within a timescale from the past to the future.

**Chapter 4: Who shapes the discourse on science and technology education?, by JD Volmink, South Africa**

This chapter identifies dominant trends or discourses in various aspects of science and technology education in African countries. These are shaped and determined by particular interest groups with conscious or unconscious agendas. The chapter examines who shapes the discourse of science and technology in Africa and analyses who and how groups, including science and technology educators, scientists and technologists, industrialists, education policy makers, economists, politicians, researchers, donors, the World Bank and foreign aid, shape discourse, practice and policy in science education.

**Chapter 5: Relevance in science and technology education, by M Rollnick, South Africa**

The importance of the relevance of the science curriculum to successful learning in science and technology education is rarely questioned. This chapter does so. Was the curriculum in the past and is the curriculum in the present relevant to the needs of Africa?

**Chapter 6: Relevance and the promotion of equity, by V Reddy, South Africa**

Historically, the participation of girls in science and technology education has been poor. In some parts of Africa certain racial groups and nomadic tribes were discriminated against, resulting in their poor participation in science and technology education. With the advent of 'science for all', equity in science and technology education has become an imperative. This chapter focuses on the challenges of access, redress, equity, and quality in science and technology education. It analyses past and present trends and proposes future directions with regard to these challenges.

**Chapter 7: Teacher education: Pre-service and in-service support models, by HM Dyasi and K Worth, USA**

The goals of science and technology education demand the implementation of good teacher development programmes. This chapter examines teacher education and support models for pre-service and in-service education used in the past and present. The authors analyse the curriculum for science teacher education; support structures such as materials, finance, and teachers' centres; relationships between schools and teacher education institutions; and teacher educators and their professional development. Importantly, this chapter delineates alternative paradigms for teacher development for the future.

**Chapter 8: Teaching large classes, by GOM Onwu, Nigeria**

After the adoption of the principle of universal primary education, the 1970s and 1980s saw an unprecedented expansion of student enrolment in African countries. As a consequence, class sizes have increased dramatically, with a concomitant decrease in the quality and quantity of resources. This chapter discusses teaching large classes in a context of poor resourcing. It examines the reality of large classes; policy and practice issues; the impact on the quality of learning in large classes; what research is available on teaching large classes; resource utilization; and innovative approaches in teaching large classes.

**Chapter 9: Resourcing science and technology education, by E Fabiano, Malawi**

The success or failure of science and technology education is dependent on the availability and utilization of appropriate resources. This chapter focuses on the quality and quantity of teachers; the role and use of print and learning materials; the impact of laboratory space, equipment and consumables on the effectiveness of practical work; the use of the school environment, and financial resources. The writer questions whether Africa can resource science and technology education on a self-sustaining basis.

**Chapter 10: The knowledge base for learning in science and technology education, by OJ Jegede, Nigeria and Australia**

An appropriate and efficacious knowledge base is paramount for science and technology learning in Africa. This chapter examines types of knowledge and ways of knowing; local cultural and indigenous knowledge systems versus the universality of Western science; second and third-language teaching of students whose mother tongue is not English; teaching classes with students of many mother tongues; cognitive styles, constructivism, and concept learning in the African child; the African child's background; the impact on learning of belonging to rural versus urban communities, and the particular cognitive problems facing girls.

**Chapter 11: Research in science and technology education, by P Naidoo, South Africa**

The main purpose of research in science and technology education is to improve policy and practice. This chapter surveys the research. Some of the issues it examines are: Who defines research? What is the African researcher's reference group? What are current research definitions and trends? Who funds and publishes research in Africa? The conspiracy of silence in research. Who is engaged in research? What assumptions direct research? Who is the proper audience for the results of research? Which are the dominant modes of research?

**Chapter 12: The mass media and science and technology education, by T Mschindi, and S Shankerdass, Kenya**

The mass media has a potentially important role to play in popularizing science and technology. This chapter focuses on modern mass media, traditional mass media, and their interface with informal and nonformal education in science and technology education.

**Chapter 13: Into the next millennium by P Naidoo, South Africa, and M Savage, Nairobi, Kenya**

This chapter attempts to synthesize the preceding chapters and summarize discussions at the ASTE '95 meeting. The synthesis focuses on the challenges and the way forward for science and technology education in Africa for the next millennium.

# Contents

# 1

# Historical perspectives and their relevance to present and future practice

*Emmanuel Ayotunde Yoloye, Professor Emeritus, Ibadan University; Director, Amoye Institute for Educational Research and Development, Ibadan, Nigeria*

## ABSTRACT

This chapter examines the historical perspectives of the last three decades and their relevance to the present and future of science and technology education. It pays particular attention to landmark meetings and organizations that had an impact on the continent. The chapter draws lessons from such organizations for the future, both at policy and at practice level.

## THE AWAKENING IN AFRICA

Political independence in Africa was an important factor contributing to the development of science and technology. Before the 1960s, most countries on the continent gave little attention to teaching these subjects. In primary schools, what passed for science was a study of nature, hygiene, health and rural science. Objectives were simple, namely the development of clean and healthy habits, an understanding of nature and the principles and techniques of farming.

In the 1950s, a few secondary schools taught physics, chemistry and biology, but their facilities and equipment were inadequate. Only two high schools in The Gambia offered science courses. In Kenya and a number of East African countries, racial considerations influenced the curriculum. Most European and many Asian schools taught science, but few African schools did. Blacks in South Africa and Namibia experienced similar discrimination. Objectives for teaching science in secondary schools were seldom stated, since teaching was geared to overseas examinations such as the Cambridge and London School Certificates.

In the early 1960s, a number of international and regional conferences drew the attention of African policy makers to the importance of science and technology education. One was the 1960 Rehovoth (Israel) Conference on Science in the Development of New States. Two recommendations of this conference were as follows:

> The Governments of developing states should regard the furtherance of science and technology as a major objective of their national politics and make appropriate provision for funds and opportunities to achieve this end ... Until such time as their own scientific manpower is adequate, new and developing states would be well advised to seek the help of scientific advisors and experts from friendly countries and international agencies to help them develop a scientific practice and tradition. (Gruber, 1961)

The 1961 Addis Ababa (Ethiopia) Conference of African States on the Development of Education in Africa, organized by the United Nations Educational, Scientific and Cultural Organization (UNESCO) and the Economic Commission for Africa (ECA), recommended that:

> African educational authorities should revise and reform the content of education in the areas of curriculum, text books and methods, so as to take account of the African environment, child development, cultural heritage and the demands of technological progress and economic development, especially industrialization. (UNESCO, 1961)

Finally, the Conference of African Ministers of Education on the Development of Higher Education in Africa was held in 1962 in Tananarive (Madagascar). The participants concluded that the ratio of students in scientific and technological fields to those in the humanities should be 60 : 40.

The Rehovoth conference drew attention to the importance of science and technology in development and the need for assistance from more developed countries. The Addis Ababa conference highlighted relevance, and identified the African environment, child development, African cultural heritage, and the demands of technological progress and economic development as four important facets of science and technology education. The Tananarive conference stressed the importance of developing local expertise in science and technology in Africa. The 60 : 40 ratio recommended in Tananarive became a guideline for university admission in many African countries. In their drive to modernize, African countries took science and technology seriously. Each country took positive steps to achieve technological and economic development through education.

## INNOVATIONS IN SCIENCE AND TECHNOLOGY IN AFRICA: A SUMMARY

### Capacity building

The first wave of curriculum reform in African countries was the development of personnel in curriculum development. This was done through initiatives such as the African Primary Science Programme (APSP) at the primary level and Nuffield science at the secondary level. Both developed and published a range of curriculum materials. In addition to this on-the-job training, both initiatives attempted to consolidate personnel development by facilitating further staff qualification at appropriate institutions within and beyond Africa. Since then, staffing at these institutions has suffered through promotion, flight to other organizations, and lack of resourcing. At the teacher level, in addition to in-service work by the various curriculum projects, donors such as the Swedish International Development Agency (SIDA) helped establish institutions such as the Kenya Science Teachers' College (KSTC).

## National projects

Having established curriculum development expertise, countries in Africa were in a position to develop a second wave of curriculum materials. These not only adapted earlier courses, but also incorporated concepts such as integrated science — especially in Nigeria — influenced by UNESCO; environmental science, influenced by the United Nations Environment Program (UNEP); and population education, influenced by the United Nations Development Program (UNDP). Many national projects, hurriedly implemented under pressure from governments and donors, were unable to involve teachers and other stakeholders and could not set up the necessary infrastructures such as teacher development programmes and appropriate examinations. Zimbabwe Science (ZIMSCI) and Botswana Science (BOTSCI) are examples of such projects. Also during this era, many countries restructured their educational systems in an attempt to make education more relevant to school leavers and to make access to higher institutions more equitable. Kenya, which in the early 1980s changed from a 7–3–2–3 cycle, with sixth-form schools as pre-university institutions, to an 8–4–4 cycle, is one example of such restructuring.

## Technical education

Technical education demands a special mention. Immediately after independence, countries such as Nigeria established secondary technical schools similar to their counterparts in the United Kingdom in an attempt to develop cadres of technologists and high-level technicians. However, due to high per student costs and the failure of graduates to find gainful employment despite loan schemes to finance their studies, these institutions were phased out. Cox-Edwards notes that in 1993 agricultural schools received 200 percent of the subsidy to general secondary schools, and industrial schools 125 percent (World Bank, 1995: 100).

In other countries, such as Kenya, similar polytechnics still function in collaboration with local industrial and manufacturing sectors. Ghana established more modest post-primary continuation schools during the early 1970s to equip students with the necessary technical skills to impact on the informal sector of the economy. These too were phased out, partially because of expense and partially because they could not compete with established, informal apprenticeship systems. Subsequent government funding policies to tertiary-level institutions to redirect their research by establishing consultancy firms in formal and informal industrial centres have been more effective in bridging academia with production; village polytechnics such as those in Kenya have been less so since village economies can only absorb limited numbers of graduates.

The history of technical education in African countries reflects current thinking by the World Bank (World Bank, 1995). Cost-effectiveness studies reportedly show that investment in technical education rarely gives higher rates of return than investment in general education.

## REGIONAL PROGRAMMES

Russia launched the first Sputnik in 1957. That historic event may have been the prime motivation for a flurry of science curriculum-development activities in the United States (US) during the late 1950s and early 1960s. Even before Sputnik, professional journals and yearbooks in the US had called for new, enlightened approaches to science teaching. The success of the Russian space programme created a sense of crisis that helped move the nation to action.

Two other events influenced science education at the time. First, an economic boom in the US made abundant funds available for domestic and international programmes. Second, new pedagogical equipment, such as film loops (these were film strips that were looped into film projectors — hence film loops — and were in use in the 1950s and 1960s), automated instructional devices, projectors and photocopiers became commonplace. The dramatic increase in foreign aid coupled with efforts in the US to renew its own national science curriculum, funded by the National Science Foundation (NSF), inevitably linked America with efforts to renew science curricula in Africa. The European Community and the United Nations also sent technical assistance in science education, for example the Nuffield science project in Britain. A regional survey carried out in 1980 (Yoloye & Bajah, 1981) mentioned 20 organizations that contributed to the development of science education in Anglophone Africa during the 1960s and 1970s. UNESCO, the United Nations Children's Fund (UNICEF) and the United Nations Development Program (UNDP) were outstanding. Their contributions included financial aid; the supply of equipment, books, teachers and experts; and training programmes for curriculum specialists and teachers. These organizations sponsored several education projects with strong science components such as the Namutamba Project in Uganda, the Mid-West (Bendel) State Primary Science Project in Nigeria and the Bunubu Project in Sierra Leone.

In many African countries, the British Council made important contributions to in-service training of science teachers, and the United States Peace Corps, the Canadian University Service Overseas (CUSO) and the British Voluntary Service Organization (VSO) provided large numbers of science teachers to secondary schools. The Swedish International Development Agency (SIDA) established the Kenya Science Teachers' College in the late 1960s for training science, mathematics and industrial-education teachers. The Canadian International Development Agency (CIDA) initiated a similar training institution for technical teachers. Other organizations that have contributed to science education in Africa include the Norwegian Agency for Development (NORAD), the Danish International Development Agency (DANIDA), the International Development Association (IDA), United States Agency for International Development (USAID), the European Economic Community (EEC), the British Overseas Development Ministry (ODM), and the Ford and Rockefeller Foundations. Perhaps the most significant intervention on a regional basis was a spin-off from the Rehovoth conference. The inspiration was provided by a Sierra Leone educator, the Reverend Solomon A Caulker, who participated in that conference. To this day, African science educators often quote Caulker. His statements include:

> The whole question, in terms of the new states, is not a question of science as a disembodied spirit, moving by itself and going into Africa. It is a question of men of science, men who will, through training, help the African people to develop. This means our schools ... To all of us has come a realization that science, through its constantly changing and growing insight, can be brought to bear to liberate the human spirit and to make us all stand with pride and believe that we are members of the human race. (Gruber, 1961)

On his return from the Rehovoth conference, Caulker died in an air crash outside Dakar. His tragic death touched Jerrold R Zacharias, an American physicist who had spearheaded the famous Physical Sciences Study Committee (PSSC) and had also been at Rehovoth. Determined to keep Caulker's spirit and ideas alive, Zacharias set up and chaired a steering committee to plan an international conference that would focus specifically on education in Africa. Funded by the Ford Foundation and the International Cooperation Administration, the African Summer Study, or Endicott House Conference, took place in Dedham, Massachusetts, in 1961. Fifteen out of the 79 participants were African.

The Endicott House Conference established the African Education Programme, funded by USAID and the Ford Foundation (EDC, 1967). As part of this effort, the African Mathematics Programme (AMP) was launched in 1961. Inspired by the School Mathematics Study Group (SMSG) in the US, the AMP produced what came to be known as 'Entebbe mathematics'. Textbooks and teachers' guides were tested in about 1 500 classrooms in Ethiopia, Ghana, Kenya, Lesotho, Liberia, Malawi, Nigeria, Sierra Leone, Tanzania and Uganda (EDC, 1967). The project introduced so-called modern mathematics to Africa, an approach that focused on teaching major, underlying conceptual structures. However, this approach soon became controversial. A number of African countries, including Nigeria and Kenya, eventually banned modern mathematics, because teachers were reported to have had problems with the approach. Nevertheless, many of the original concepts persist in present-day curricula throughout Africa.

Following the Endicott House Conference, the Ford Foundation funded experimental projects in Kenya and Nigeria. In Kenya, a science centre undertook science curriculum development, the production of classroom science equipment, and the training of primary science teachers. In Nigeria, Babs Fafunwa, who had been at

Endicott House, organized a series of workshops in primary school science at the University of Nigeria, Nsukka. Mike Savage, who had also been at Endicott House and had subsequently participated in the Elementary Science Study (ESS) in the US, worked through the University of Nigeria with primary schools in nearby Awo Omama.

In February 1964, a conference was held in Kano, Nigeria, that marked the formal launching of the African Primary Science Programme (APSP). Babs Fafunwa from Nigeria, John Gitau from Kenya, Ron Wastnedge of the Nuffield Junior Science Project in the UK, Len Sealey of the Leicestershire Education Department in the UK and Phil Morrison of Cornell University in the US presented their experience with innovative science education projects. Mike Savage worked for two weeks with a group of primary school teachers from Kano, and these teachers gave demonstration lessons that persuaded participants that an inquiry approach to science teaching was effective with teachers and pupils in Africa.

Under the guidance of Jack Goldstein, an astrophysicist at Brandeis University, participants from Africa, the US and the UK developed classroom materials at three regional workshops. These were held in Entebbe, Uganda (1965), Dar es Salaam, Tanzania (1966), and Akosombo, Ghana (1967). APSP helped create science centres in Ghana, Kenya, Malawi, Nigeria, Sierra Leone, Tanzania and Uganda. Science educators in these centres worked for several years in classrooms trying out materials and modifying them in the light of experience. The project produced more than 30 units and eight background readers.

With the creation of the Science Education Programme for Africa (SEPA) in 1970, APSP management passed into African hands. Hubert Dyasi, SEPA's first executive secretary, established the secretariat in Accra, Ghana. SEPA programmes were established in Botswana, Ethiopia, The Gambia, Ghana, Kenya, Liberia, Lesotho, Malawi, Nigeria, Sierra Leone, Swaziland, Tanzania, Uganda, Zambia and Zimbabwe. Unfortunately, SEPA collapsed in 1985, primarily due to a lack of external funding. However, this programme had a profound influence on science education in many African countries that is still in evidence today. I shall discuss the legacy of SEPA later in this chapter.

During the early 1970s, UNESCO organized a nine-month workshop in integrated science for African curriculum development specialists. This influential workshop, which took place at Cape Coast, Ghana, spearheaded integrated science teaching in many African countries. Integrated science became particularly rooted in Nigeria where the Science Teachers' Association of Nigeria (STAN) ran a series of writing workshops. Schools all over Nigeria have adopted the approach and teaching materials introduced by this project.

Finally, the Centre for Development Cooperation of the Free University of Amsterdam, in the Netherlands, collaborates with universities in Botswana, Lesotho, Mozambique, Malawi, Namibia and Swaziland to increase the number of science undergraduates through bridging and remedial courses. The centre has introduced innovative models of in-service teacher development.

## IN-COUNTRY PROJECTS

In addition to these regional programmes, many African countries had their own science projects, often with support from external sources.

### The Namutamba Project (Uganda)

In 1967, the Ugandan government established the Namutamba Project with UNESCO support. The project's aim was 'to improve living conditions in a selected rural area and to assist the children, youth and adults to prepare for effective and rapid integration into the social, cultural and economic development of Uganda'.

The project developed a functional rural science curriculum, a rurally oriented primary education programme and a comprehensive formal and nonformal education programme for rural development. Tutors and trainees of Namutamba Teacher Training College developed innovative primary-level curriculum materials, and these were introduced in 15 primary schools associated with the project. In an evaluation commissioned by SEPA and UNESCO, the International Centre for Educational Evaluation (ICEE) at the University of Ibadan, Nigeria, found that the project had succeeded in changing the attitudes of teachers and pupils towards agricultural occupations, rural studies and living in rural areas (Yoloye & Bajah, 1975).

### The Bunubu Project (Sierra Leone)

The Bunubu Project in Sierra Leone began in 1974 with support from UNESCO and UNDP. It was similar to the Namutamba project. Located in a rural area at Bunubu Teachers' College, the project was associated with 20 primary schools. Its aim was 'to improve the quality of life in rural areas through the medium of education'. The project provided primary education with a rural bias, trained primary school teachers in community development, and implemented community development and adult education programmes. The project added agricultural science, home economics, practical arts, community development and adult education to science in the regular curriculum.

A unique feature of the Bunubu project was the close involvement of community chiefs and other leaders. Project staff explained their philosophy to local leaders, who helped form community development councils that closely interacted with the college and associated schools. Local artisans taught in the college and schools, and teachers and pupils organized adult education programmes in the community. Together, the schools and the community organized projects such as fish ponds and cash-crop farming. An in-depth evaluation found that this type of community/school interaction led to increased community development efforts and to changes in attitudes, both within the community and in the schools (Lucas, Yoloye & Sissay, 1987).

### The Mid-West State Primary Science Project (Nigeria)

In 1968, the Mid-West (Bendel) State Government of Nigeria established the Mid-West State Primary Science Project (MPSP) with assistance from UNESCO and UNICEF. The project established an in-service training centre at Abraka to train primary school teachers and to develop science curriculum materials that included student texts and teachers' guides.

The project operated in 100 pilot schools for six years and was then implemented on a state-wide basis. Longmans (Nigeria) Ltd published a series of books, entitled *Science is Discovering*, that aimed at developing in children an attitude of inquiry, plus observing, exploring, experimenting and recording skills, and an understanding of the basic concepts of cause and effect.

In 1976, ICEE evaluated the project at the request of UNICEF (Falayajo, Bajah & Yoloye, 1976). The evaluation report indicated that the project had had a favourable impact on teacher performance. The impact on the pupils was more difficult to measure: because many nonpilot schools had teachers trained under the project, there was no proper control group.

### Zimbabwe Science and Botswana Science

Zimbabwe Science (ZIMSCI) was based on an inexpensive science kit designed for secondary school leavers. Conceived as a means of distance science teaching, ZIMSCI was intended to function independently of teachers. This proved to be a weakness, and the project suffered from inadequate support to teachers. With greater financial support, Botswana Science (BOTSCI) was a school-based programme adapted from ZIMSCI. The BOTSCI science kit included glassware and various chemicals, while the ZIMSCI kit used inexpensive equipment such as milk tins and bottles. Crash training programmes in Botswana converted humanities teachers to science teachers, and expatriate teachers were also hired under the project.

### The Science Education Project (South Africa)

The Science Education Project (SEP), started in 1976, is one of many innovative projects in South Africa (Kahn & Rollnick, 1993). SEP uses low-cost, locally manufactured equipment. Unlike ZIMSCI and BOTSCI, the project is geared to an existing syllabus. Most rural areas have adopted SEP, but the project scarcely exists in urban areas. Reportedly, only 50 % of white schoolchildren and 17 % of black schoolchildren study science in South Africa, and only 5 % of black teachers are qualified to teach physical science. The situation in Namibia is in many respects similar to that in South Africa. Recent political changes in these countries have provided a fresh impetus to innovate in science teaching.

## THEORETICAL AND PHILOSOPHICAL CONTEXT

A long history of theoretical and philosophical thinking about science teaching, primarily in Europe and the US, has influenced teaching in Africa. Conversely, science teaching in Africa has made contributions to thinking elsewhere. To illustrate this process of dialogue I shall consider the basic curriculum questions — Whom to teach? What to teach? How to teach?

## Whom to teach?

Although modern experimental science emerged in the 16th century, science teaching became part of the curricula of formal educational institutions only slowly. According to Lauwerys (1957):

> Science had been given its head in industry but had been frustrated and hamstrung in education. In so far as scientific knowledge was evidently essential to the then modern living, it was provided within industry itself or in special institutions called 'technical colleges' which were regarded as inferior institutions and seldom attracted the high caliber or the upper classes.

It was not until the late 19th century that science became part of the school curriculum in the US and continental Europe. In England and Wales, it was not until the early 20th century. As the impact of science and technology on economic development, and on society generally, has become more evident, courses on science and technology have become more common. In the 1980s and 1990s, this trend has broadened into an advocacy of 'science for all', sometimes called the 'scientific literacy' movement.

Thus, over the years, the answer to the question, 'To whom shall we teach science?', has changed from a few low-grade technicians, to the students in formal education institutions, to all citizens. This trend can be seen in many Anglophone African countries. During the colonial era, access to schooling was limited for Africans, with the primary aim of producing low-level technicians. Immediately before and after political independence, education for Africans became more elitist, reflecting the need to replace Europeans in upper-level and middle-level technical and management positions. This was soon achieved.

Political pressures then led to a rapid expansion of educational opportunities, especially at the primary levels. With the dramatic expansion of access to education, the content became increasingly pre-vocational, rather than merely preparing pupils for the next stage of schooling. The aim was to equip school-leavers to lead constructive lives in the non-formal rural and urban economies. Technical secondary schools in Nigeria, continuation schools in Ghana, and the village polytechnics in Kenya were established during this era. At first, the more formal, main-stream schools were still perceived as leading to salaried positions in the formal sector.

With continued expansion, the emphasis has changed even in mainstream schools. During the early 1980s, Kenya changed from a national system of seven years of primary, six years of secondary and three years of tertiary (7–6–3) education to an 8–4–4 system. The longer primary cycle was designed to provide children with appropriate life skills, and the shorter secondary cycle opened access to an expanded university system. Such far-reaching changes in educational systems throughout Anglophone

> Africa have put strains on national and household budgets. They have also generally been achieved at the cost of a loss of quality, and they have led to large numbers of school-leavers without salaried jobs. However, increases in school enrolment have been remarkable.
>
> One important issue throughout the expansion of educational opportunities in Africa has been the under-representation of girls in the sciences. In the past two decades, gender in science teaching has assumed global importance. Under-representation of women is partly rooted in the history of the development of science. The modern scientific method emphasizes logical reasoning and an assumption that natural phenomena have rational explanations. The common belief in the 16th century was that men were ruled by reason and women by emotions (Harding, 1992). Women were seen as unsuited to the study of science, and thus most pioneers of modern science were men. There has therefore been a dearth of female role models and inadequate opportunities for girls to study science and technology.

Harding (1992), Awe and Adedeji (1990), and the African Academy of Sciences (1995) studied the factors leading to gender imbalance in science, technology and mathematics. Considering the findings of such research, developed countries and some African countries have intervened in the educational process to reduce such imbalances. Intervention strategies have included:

- Introduction of legislation to promote equal opportunities for men and women in science, technology and mathematics education and careers.
- Support for special training programmes to facilitate the entry of women into science and technology careers.
- Change from predominantly single-sex to mixed-sex schools.
- Development of mobilization and enlightenment programmes.
- Policies to make mathematics and at least one science subject compulsory in secondary schools.
- Modification of science, technology and mathematics curricula to make them nonsexist.
- Organization of training programmes for women workers in nontechnology fields so they can move into technology-related jobs.

Efforts to correct gender imbalance in science, technology and mathematics education are gathering momentum. In particular, the Donors to African Education (DAE) show keen interest in this area.

## What to teach?

Answering the question 'Whom to teach?' raises another question: 'What to teach?' Because the range of 'whom' is so diverse, 'what' is taught must also vary according to the learners' educational backgrounds, abilities, and goals.

Until the 20th century, the goals of a society and the organized body of knowledge available were the primary factors influencing the content of education. With a rise of studies in human development and learning psychology, the emphasis in curriculum planning shifted so that the nature of pupils and their learning processes assumed greater importance in choosing what to teach (NEA, 1963).

After World War II, there was a new emphasis on science and technology and a significant expansion and proliferation of scientific disciplines. Educators sought ways to help pupils learn science as quickly as possible. Many science curriculum projects in the US reflect this emphasis. In 1959, the National Academy of Sciences (NAS) sponsored a 10-day conference at Woods Hole, Massachusetts, that significantly influenced the direction of science and mathematics curriculum development in America. Participants included 16 scholars in science and mathematics, 10 in psychology, and three each in the humanities, education and cinematography. They discussed new educational methods, particularly in science. In a summary of these discussions entitled *The Process of Education*, Jerome Bruner (1960) identified four important elements of curriculum development:

1. **The structure of knowledge**: 'Grasping the structure of a subject is understanding it in a way that permits many other things to be related to it meaningfully. To learn structure, in short, is to learn how things are related.'
2. **Readiness for learning**: 'We begin with the hypothesis that any subject can be taught effectively in some intellectually honest form to any child at any stage of development.'
3. **Intuition in learning**: Participants defined intuition as 'the intellectual technique of arriving at plausible but tentative formulations without going through the analytic steps by which some formulations would be found to be valid or invalid conclusions'. They believed that scientific intuition plays a crucial role in the advancement of science.
4. **Motivation**: Learning depends on the desire to learn. Participants agreed that interest in the material to be learned is the best stimulus to learning, rather than external goals such as grades. However, they thought that much can be done to provide intrinsic motivation by a manipulation of the learning climate in the school and attitudes within the community.

These four elements provide the basis for my discussion on the content of science and technology curricula: 'What to teach?'

### *The structure of knowledge*

In the 1960s, many science and mathematics curriculum projects in the US emphasized structure. At the primary level, 'Science, a Process Approach' (SAPA) focused on the processes of science such as observing, using space/time relationship and numbers, measuring and classifying. The 'Science Curriculum Improvement Study' (SCIS) identified scientific concepts such as material objects, interactions, systems and subsystems, relativity, organisms and life cycles. At the secondary level, a notable example was the 'Chemical Bond Approach' (CBA), launched in 1959. Reasoning that the making and breaking of bonds is at the heart of chemical change and chemistry, CBA built a curriculum around the central theme of chemical bonds.

Trials, however, showed that concentrating on chemical bonds made chemistry too abstract. Besides, chemical bonding is just one conceptual model to explain chemical reactions. The final version of this project, entitled 'The Use of Conceptual Models in Explaining the Behaviour of Chemical Systems', used a variety of models to explain chemical reactions.

A major problem of the 'structure' approach to curriculum development — as the Entebbe mathematics project experienced — is that any subject has more than one structure. Another problem is that courses based on structure tend to be abstract. In Africa, this approach was unfamiliar to parents, teachers found it difficult, and political leaders gained support by opposing it as a 'foreign' import.

### *Readiness for learning*

Bruner based his hypothesis on 'readiness for learning' on the experiments of the developmental psychologist Jean Piaget. Piaget's work made it clear that children begin to grasp concrete operations at about the age of seven, the age when they normally begin primary school. At this age, children can learn fairly sophisticated scientific concepts, provided materials are used and teaching focuses on the concrete, operational level. Based on this hypothesis, many science and mathematics curricula, such as SMSG mathematics and its derivative 'Entebbe mathematics', taught concepts in primary and secondary schools that were previously taught only at the university level.

### *Intuition in learning*

In an effort to improve understanding of 'scientific intuition', Marton, Fensham and Chaiklin (1994) analysed discussions with 93 Nobel prize winners in physics, chemistry and medicine. Seventy-two of these researchers believed in scientific intuition. The authors summarized the Nobel laureates' views as follows:

> Scientific intuition is seen as an alternative to step by step logic and is closely associated with a sense of direction. It is more often about finding a path than arriving at an answer or reaching a goal ... Intuition is rooted in extended, varied experience of the object of research. Although it may feel as though it comes out of the blue, it does not come out of the blue.

One dilemma of science education is whether to characterize intuition as part of the so-called scientific method. For centuries, *Organon*, a collection of Aristotle's treatises on logic, provided the acknowledged basis for the study of natural science. In the 13th century, Roger Bacon investigated nature using techniques other than logic. He and others like him, however, tended to be regarded as wizards in league with evil spirits, partially because in those days experimental science was represented by alchemists who tried to transmute baser metals into gold and cloaked their operations in mystery.

It was not until the 16th century, in the latter part of the Renaissance, that modern experimental science began to emerge. Francis Bacon (1561–1626) played a key role, publishing the *Novum Organum* or 'New Instrument' to replace Aristotle's *Organon.* He called his method 'true induction'. Bacon himself made practically no contribution to scientific knowledge, but his advocacy of basing investigation on facts and experimentation strongly influenced his contemporaries. These included William Gilbert (1578–1603), the founder of the sciences of electricity and magnetism, and William Harvey (1578–1657), who discovered the circulation of the blood. The astronomer Galileo Galilei (1564–1642), another contemporary of Bacon's, used the scientific method frequently and contributed to the development of science by his recognition of the role of hypothesis and mathematical reasoning. In describing the development of the scientific method, Margenau and Bergamini (1964) write:

> The term scientific method itself was something of a misnomer. It is not a method in the sense of a final procedure. It furnishes no detailed map for exploring the unknown, no surefire prescription for discovery. It is rather an attitude and philosophy, providing guidance by which dependable overall concepts can be extracted from impressions that swarm in on man's senses from the outside world . . . With its virtues, the method has certain limitations. It cannot replace the inspiration of Archimedes discovering a basic law of hydrostatics while sitting in his bath. It cannot conjure up the good luck of Alexander Fleming chancing on penicillin. It cannot hasten the slow process of intellectual growth and reasoning. In short, it cannot create science automatically any more than the theory of harmony can write a symphony, or a naval manual can win a sea battle.

Such views notwithstanding, experience with the scientific method is likely to prepare an individual to profit from an Archimedian-type inspiration or a Fleming-like stroke of luck. The journal *Chemistry,* published by the American Chemical Society, printed a series of articles in 1966 called 'Chance favors the prepared mind'. The series dealt with accidental discoveries such as the first synthetic dye, mauve, by William Henry Perkin in 1856, when he was just 17 years old, and dynamite by Alfred Banhard Nobel in 1867. Although many significant discoveries are made by chance, the authors of the series emphasized that it takes people with certain skills, attitudes and philosophies to capitalize on such chances or accidents. The teaching of these skills, attitudes and philosophies is an essential element of many science curricula.

Duckworth (1978) suggests her own solution to the dilemma of whether to characterize intuition as part of the scientific method in the chapter headed 'The having of wonderful ideas' (1978: 18–28). Wonderful ideas are flashes of inspiration or insight, intuitive ways of tackling identified problems. Here are some of her observations:

- 'The having of wonderful ideas is what I consider the essence of intellectual development' (1978: 18).
- 'Wonderful ideas do not spring out of nothing; they build on a foundation of other ideas' (1978: 23).

- 'Wonderful ideas are built on other wonderful ideas. They do not occur contentless. In Piaget's terms, you must reach out to the world with your own intellectual tools and grasp it; assimilate it yourself' (1978: 24).
- 'If a person has some knowledge at his disposal, he can try to make sense of new experiences and new information related to it. He fits it into what he has. By knowledge I do not mean verbal summaries of somebody else's knowledge. I mean a person's own repertoire of thoughts and actions, connections, predictions and feelings. Some of these may have as their source something he has read or heard. But he has done the work of putting them together' (1978: 27).
- 'The more ideas a person already has at his disposal about something, the more new ideas occur and the more he can coordinate to build up still more complicated schemes' (1978: 28).

Mike Savage (1994) equates wonderful ideas with creativity and insight. He considers both indispensable to scientific education. A consensus would be that scientific intuition is most likely to develop when a pupil is exposed to diverse experiences with relevant materials. The provision of such experiences therefore constitutes an essential part of a good science curriculum.

### *Motivation*

Child-oriented science projects such as ESS and APSP were based on a belief that children could be motivated to have an intrinsic interest in learning through the use of materials or problems. This belief led curriculum-development specialists to work with children to find out their interests and to use these interests as the basis of teaching units. This approach also implies that no single set of materials can be used to teach science to all children in all situations. Curriculum developers must work with children to determine what approaches and materials will provide a basis for successful teaching.

## How to teach?

Over the years, science teaching has moved from rote learning to an emphasis on learning for understanding. During the curriculum innovations of the 1960s, an emphasis on inquiry, discovery, and problem solving became prominent. This emphasis was largely a by-product of the then-current focus on the processes of science. It implies a strategy for developing understanding. Some scientists advocate a focus on process as the essence of science education. Sir James Jeans (1958) wrote: 'To many, it is not knowledge but the quest for knowledge that gives the greatest interest to thought ... To travel hopefully is better than to arrive.'

Hawkins (1965) identifies three phases in the inquiry process. He calls the first period 'messing about', when children are encouraged to explore, manipulate and try out ideas with materials and equipment. This period may be extended over weeks if interest is high. Second is a phase of directed, individual investigation. The third phase involves pooling information, discussing ideas, and extracting generalizations.

Hawkins based his advocacy on his experience with ESS. In Africa, much of the inspiration for APSP came from ESS, and APSP also adopted this three-part procedure to teaching primary science. Here it is important to make a distinction between at least two levels of inquiry. One is free inquiry, or 'messing about', when children identify and solve their own problems. Another is often called guided inquiry, when well-sequenced investigations lead children to predetermined knowledge. Whenever students work to a set syllabus, there is a preference for guided inquiry. As defined by Hawkins, however, the guided-inquiry phase does not lead necessarily to predetermined knowledge, but rather to the solution of problems identified by the students. In this sense, it is an extension of free inquiry. Both APSP and ESS abound in examples of this approach.

Attractive as this inquiry/discovery/problem-solving approach was, it was not without controversy. Bruner (1960), for example, states:

> Intellectual activity anywhere is the same whether at the frontiers of knowledge or in a third grade classroom ... The difference is in degree, not in kind. The schoolboy learning physics is a physicist and it is easier for him to learn physics behaving like a physicist than doing something else.

Ausubel (1969) is of a different opinion:

> First, I cannot agree that the goals of the research scientist and that of a science student are identical ... Thus while it makes perfectly good sense for the scientist to work full time formulating and testing hypotheses, it is quite indefensible in my opinion for the student to be doing the same thing — either for real, or in the sense of rediscovery.

In the last decade, a variant of the inquiry/discovery/problem-solving paradigm has been widely advocated and studied under the label of 'constructivism'. Different authors have described constructivism as follows:

- Constructivism is an epistemology that focuses on the role of learners in the personal construction of knowledge. (Ritchie, 1994)
- Learning is viewed as an adaptive process where existing knowledge is modified in response to perturbations that arise from personal and social interactions. (Ritchie, 1994)
- In a constructivist classroom, students are encouraged to take responsibility for their own learning as they explore. (Ritchie, 1994)
- In class, students try to make sense of experiences in terms of their prior knowledge.
- Active teaching is required to monitor student understanding and help them restructure ideas through negotiating meaning. (Driver, 1988)

Studies on constructivism abound in science education journals. Examples are those of Baimba, Katterns and Kirkwood (1993), Gaskell (1992), Watts and Bentley (1991), Tobin (1990), Harlen (1992), and Marin and Benarroch (1994). Constructivism has become central in educational research. Magoon (1977) labelled as constructivist approaches research techniques that have been variously referred to as anthropological, participant/observer, phenomenological, ethnographic and humanist.

It is interesting to compare Ritchie's characterization of constructivism with Duckworth's 'The having of wonderful ideas', which she conceptualized after her experience with APSP in the 1960s. I believe the Africa Primary Science Programme, and Duckworth herself, were constructivist before the term was used in science education. Actually, I think they were more than constructivist as this term is currently defined in the literature. For this reason, I shall discuss APSP in more detail.

## APSP AND SEPA

The African Primary Science Programme (APSP) was unique in that it did not bother with labels. It had only one goal, namely to help children do and learn science. Curriculum development specialists who were closely connected with APSP describe this uniqueness as follows:

> The African Primary Science Programme shared with the Elementary Science Study the tendency, among other things, to leap into the fray without starting from a detailed statement of goals and objectives. (Duckworth, 1978)

> There appeared to be a remarkable reluctance, or was it inability, on the part of these people to verbalize what they were trying to do. Yet there was little doubt that they were doing something promising and exciting. (Yoloye, 1978)

Evaluators led by Yoloye and Duckworth compiled goals for APSP three years after the project started. These were based on observation of what was happening in classrooms. Many science educators described the approach as inquiry/discovery/problem-solving. Yoloye (1978) characterized APSP teachers as 'open' and characterized the programme as 'humanist' (Yoloye, 1994). Perhaps no single label completely captured the programme's spirit. How do we explain how a primary school science unit called 'Ask the Antlion' so intrigued an experienced teacher that she kept investigating for nine months? Listen to her 'wonderful idea':

> His [Yoloye's] approach ... generated in me the desire to study the antlion beyond any study undertaken by others in my class, and finally perhaps to lead me to some contribution in the study of nature in my immediate environment ... I had been successful at keeping an antlion alive for three whole weeks, an achievement which was not recorded in any book I have so far read. (Ayankogbe, 1978)

Mrs Ayankogbe had reared several antlions from larva to adulthood and had hoped the adults would mate and produce eggs so that she could document the entire life cycle. She had had no formal science training before joining a one-year diploma class where she was introduced to APSP materials.

APSP and its successor, the Science Education Programme for Africa (SEPA), are perhaps best characterized by the influence that they still exert on science education in Africa. I shall examine how these programmes have affected science education by influencing classroom practice, developing institutional linkages, building human resources, conducting research and ensuring sustainability.

## Influencing classroom practice

APSP and SEPA have had a significant impact on the philosophy and practice of science teaching in Anglophone African countries. It is interesting to speculate why these programmes had a greater long-term impact than many others. Unlike the mathematics and secondary school science curriculum projects that were initiated at about the same time, APSP and SEPA centred on the child rather than on the structure of an academic discipline. This approach demanded little new content knowledge of teachers or parents. Learning began with children's exposure to local materials rather than with abstractions. Although this may have been a novel approach in schools, it was a common approach to learning in African societies, familiar to both parents and teachers. Changes in children's behaviour — their ability to manipulate materials and to explain their investigations in everyday language — were easily recognizable indicators of effective teaching. The relevance to community life of what children learned was also clear. To borrow from John Volmink, APSP and SEPA involved all stakeholders in the discourse not only on science education, but on education in general, empowering teachers, parents and children. During the 1970s, significant numbers of teachers could be found using the approach in classrooms and training colleges throughout Africa, as Savage has documented. Although rising school enrolments and deteriorating economies have made it increasingly difficult to implement the APSP/SEPA approach, it remains an ideal for which to strive.

## Developing institutional linkages

If we take the Rehovoth conference as a beginning, the institutional life span of APSP and SEPA was about 25 years (1960–1985). This gives some idea of how long it takes for an innovative programme to become established. There is little doubt that by 1980 SEPA had become a force to be reckoned with, both regionally and internationally. Although, for reasons mentioned earlier, SEPA lapsed into dormancy around 1985, some of the structures and institutions it established, the human resources it developed, and the vision it advocated continue to make positive contributions to science education in Africa.

APSP was initially a programme of the US-based Education Development Center (EDC) and later evolved into a programme of an independent African organization, SEPA. It started with a focus on only two countries, Nigeria and Kenya, which gradually expanded to seven countries and then to 15.

The transition from being a US-based to an independent African programme carried important lessons. In reality, a combination of independence, dependence and interdependence came into play. In the governance of SEPA, the African member countries formulated policy through a representative council. Member countries paid annual dues, but SEPA still depended on donor funds for many of its programmes. Programmes and policies arose from a cross-fertilization of ideas and experiences from more than 15 African countries.

Most significantly, SEPA embarked on a strategy of mobilizing African personnel to help individual countries on specific projects. To do this, the programme established links with several regional and international organizations, including UNESCO, UNICEF, the United Nations Environmental Program (UNEP), ODA, BREDA, the African Curriculum Organization (ACO) and African Bureau for Educational Sciences (BASE). One result of these broad linkages was the development of a large, diverse group of stakeholders in the project. These included teachers, science educators, scientists, psychologists, regional and international nongovernmental organizations (NGOs), professional teachers' associations and consultants. Such a broad diversity of stakeholders was a major source of strength for an organization hoping to carry out sustainable changes in systems of education.

## Building human resources

The experience of developing and testing APSP materials affected many teachers, scientists, science educators and ministry officials, and they in turn transferred their new skills to their colleagues. Human resource development was significantly expanded under SEPA through the establishment of the International Centre for Educational Evaluation (ICEE) at the Institute of Education, University of Ibadan, Nigeria, and the Science Teacher Educators' Programme (SETC) at the Science Curriculum Development Centre, Njala University College, University of Sierra Leone.

Established in 1972 under this author's directorship, ICEE trained educational evaluators at postgraduate diploma, master's and doctoral levels. Students went on to assume high-level positions in their home countries. SETC was established in 1975 under the directorship of Alieu Kamara. This programme trained science educators at the diploma level to become classroom teachers, ministry staff, and faculty of teacher training colleges.

Between 1972 and 1980, ICEE trained 124 students from 17 African countries — 63 at the diploma level, 53 at the master's level and 8 at the doctoral level. In the process, faculty and students conducted a great deal of fundamental educational research. In 1976, ICEE played an important role in establishing an influential NGO, the ACO, that brought together national curriculum development centres from 19 African countries. Between 1980 and 1986, ACO sponsored 32 students from member countries for master's programmes at ICEE. Funding for this programme was provided by the German Foundation for International Development (DSE). Other students came to ICEE with support from AMP, the Carnegie Corporation, CFTC, the Deutscher Akademischer Austauschdienst (DAAD), the Ford Foundation, Makerere University, and SEPA. Today, ICEE is an established department of the University of

 

Ibadan that produces about 30 graduates a year. The centre's geographical coverage has been reduced, however, due to a lack of external funding.

## Conducting research

Primary science curriculum development, as undertaken by APSP, was an effective form of action research, involving science educators, trial teachers and schoolchildren. The APSP approach took into account the three elements of relevance listed in the 1960 Addis Ababa declaration, namely the African environment, child development and cultural heritage. The programme developed teachers and educators across the continent with valuable experience in relevant action research. Even today, these individuals form a powerful reservoir from which to draw new initiatives.

SEPA carried the research thrust further by initiating basic research on the intellectual development of African children. With funds from UNEP, SEPA set up a task force that brought together research results from all over the continent. Their work resulted in a monograph entitled *The Child in the African Environment,* edited by Romanus Ohuche and Barnabas Otaala. A third contribution was the research conducted over the years at ICEE. Graduates from ICEE programmes are found today in African universities, colleges of education and curriculum development centres, providing leadership in educational research and evaluation.

## Ensuring sustainability

In an effort to ensure sustainability, the founders of SEPA worked to institutionalize specific programmes, such as ICEE and SETC, that were integrated into national university systems. Less successful was the institutionalization of SEPA itself. As an intergovernmental organization, SEPA established its legal status through an agreement with the government of Ghana and obtained observer status in the Organization of African Unity (OAU). Thus the programme achieved legal sustainability and, as a legal entity, is still alive today. During its early years of expansion, the success of SEPA was due in large part to the creativity, vision and diplomatic skills of its first executive director, Hubert Dyasi, as well as the drive and commitment of the country representatives on the programme's executive council. Unfortunately, the subsequent leadership of SEPA was not as strong, and lapses in management resulted in a loss of funding. Today the programme is dormant. Several lessons can be learned from this experience:

- For long-term sustainability, organizations need to move from dependency to interdependence in their relationships with donor agencies. Dependency hinders the development of self-reliance that forms the basis of genuine interdependence.
- Organizations need excellent leadership on a sustained basis, leadership that combines management and diplomacy skills in addition to expertise in science education.

- Donor–organization relationships are built on trust. Officers of organizations seeking donor support must ensure that the basis for trust is never eroded.
- Although a functioning secretariat is indispensable, the critical indicator of the success of an education programme in Africa is its projects and activities rather than a complex administrative structure. Thus, SEPA's activities continue to exert a strong influence on science education in Africa in spite of its demise as a formally functioning programme.

## THE FUTURE

Because of economic deterioration and a massive exodus of talented personnel, the quality of science education has declined drastically in most African countries. Despite the tremendous efforts of programmes such as APSP/SEPA, the gap in science education between the developed world and Africa widens. Books and equipment are obsolete and in bad repair, scholarly journals are unavailable, and there are few opportunities for African science educators to interact with their counterparts in other parts of the world. At a more profound level, questions are being raised as to whether the African context is conducive to the promotion of quality science education.

The African Forum for Children's Literacy in Science and Technology (AFCLIST), launched in 1988 as an activity of the Rockefeller Foundation, shows promise for the future. AFCLIST is an informal association of African educators, scientists, technologists, media specialists and international resource people. It operates a small grants programme to support innovative science education in African.

AFCLIST is a legacy of APSP/SEPA. Philosophies are similar, and many veterans of APSP/SEPA are actively involved in AFCLIST at both administrative and field levels. AFCLIST has some features that are unique in today's environment and may provide guidelines for the future. For one thing, AFCLIST primarily supports initiatives arising from African countries or from consortiums of African science educators — a policy that is most likely to ensure relevance, commitment and sustainability.

In the face of a gloomy situation, African teachers and educators must continue to strive for excellence in science education. The experiences described in this paper provide some suggestions for the future:

- Science education programmes in Africa still require funding from donor agencies, but they need to move towards interdependence rather than dependency.
- To derive optimum results from external aid, policy makers in science education must clearly identify their needs and order their priorities. Funded programmes should originate from their intended beneficiaries.
- Science education programmes require a long period of gestation if they are to engender sustainable change in education systems: planners need to adopt a long-term approach.
- In view of scarce human resources, networking should be vigorously pursued through regular communication, exchanges, collaborative research and joint action.

- Tested approaches to curriculum change can be successful in primary and secondary schools. The range of actors needs to be broad, including NGOs, the private sector, teachers' associations and institutions of higher learning.

## REFERENCES

African Academy of Sciences. 1995. *Directory of Researchers on Female Education.* Nairobi: Academy Science Publishers

Ausubel, DP. 1969. Some psychological and educational limitations of learning by discovery. In HO Anderson (ed). *Readings in Science Education for the Secondary School.* New York: Macmillan, p 108

Awe, B & Adedeji, P. 1990. *Girls and Women Education in Nigeria: A Seminar on Girls' Education in Nigeria, Primary and Secondary.* Ibadan: Institute of African Studies

Ayankogbe, A. 1978. Investigations with the antlion. In *Handbook for Teachers of Science.* Accra: SEPA, pp 8–12

Baimba, P, Katterns, R & Kirkwood, V. 1993. Innovation in a science curriculum: A Sierra Leone case study. *International Journal of Science Education,* 15(3), pp 213–19

Bruner, JS. 1960. *The Process of Education.* Cambridge: Harvard University Press

CBA (Chemical Bond Approach). 1963. *Chemical systems.* New York: McGraw Hill

Duckworth, Eleanor. 1978. *The African Primary Science Programme: An Evaluation and Extended Thoughts.* Grand Forks: North Dakota Study Group in Evaluation

Driver, R. 1988. Theory into practice II: A constructivist approach to curriculum development. In PJ Fensham (ed). *Development and Dilemmas in Science Education.* London: Falmer Rees, pp 133–49

EDC (Educational Development Center). 1967. *A Report of an African Education Program.* Newton, Ma: EDC

Falayajo, W, Bajah, ST & Yoloye, EA. 1976. *Mid-West (Bendel) State Primary Science Project.* ICEE Evaluation Report No 2. Ibadan: ICEE, University of Ibadan

Gaskell, PJ. 1992. Authentic science and school science. *International Journal of Science Education,* 14(3), pp 265–72

Gruber, R. 1961. *Science and the New Nations.* New York: Pyramid Books

Harding, J. 1992. *Breaking the Barrier: Girls in Science Education.* Paris: IIEP

Harlen, W. 1992. Research and the development of science in the primary school. *International Journal of Science Education,* 1(5), pp 491–503

Hawkins, D. 1965. Messing about in science. *Science and Children,* pp 25–9

Jeans, J. 1958. *Physics and Philosophy.* Ann Arbor, Mich: University of Michigan Press, p 2

Kahn, M & Rollnick, M. 1993. Science education in the new South Africa: Reflections and visions. *International Journal of Science Education,* 15(3), pp 251–72

Lauwerys, JA. 1957. Scientific humanism. In Judges, AV (ed). *Education and the Philosophic Mind.* London: George G Harrap

Lucas, G, Yoloye, EA & Sissay, S. 1987. *Republic of Sierra Leone: Dissemination of Innovative Primary Education Curriculum.* SIL/85/009, evaluation report. Freetown: UNDP

Magoon, AJ. 1977. Constructivist approaches in educational research. *Review of Educational Research,* 47(4)

Margenau, A & Bergamini, D. 1964. *The Scientist.* New York: Time Incorporated, Life Science Library, pp 51–2

Marin, N & Benarroch, A. 1994. A comparative study of Piagetian and constructivist work on concepts in science. *International Journal of Science Education,* 16(1), p 115

Martin, Fensham, P & Chaiklin, S. 1994. A Nobel's eye view of scientific intuition: Discussions with Nobel prize winners in physics, chemistry and medicine. *International Journal of Science Education,* 16(4), pp 457–74

NEA (National Education Association). 1963. *Deciding What to Teach.* New York: McGraw Hill, pp 10–18

Ritchie, SM. 1994. Metaphor as a tool for constructivist science teaching. *International Journal of Science Education,* 16(3), pp 293–304

Savage, MBR. 1994. The having of wonderful ideas. *The African Forum for Children's Literacy in Science and Technology Newsletter.* July, pp 1–6

SEPA (Science Education Programme for Africa). 1978. *Handbook for Teachers of Science.* Accra: SEPA, pp 82–3

Tobin, K. 1990. Social constructivist perspectives in the reform of science education. *Australian Science Teachers Journal,* 36(4), pp 29–35

UNESCO (United Nations Educational, Scientific and Cultural Organization). 1961. *Conference of African States on the Development of Education in Africa: Final Report.* 15–23 May, Addis Ababa. Paris: UNESCO

Watts, M & Bentley, D. 1994. Humanizing and feminizing school science: Reviving anthropomorphic and animistic thinking in constructivist science education. *International Journal of Science Education,* 75(1), pp 83–9

Yoloye, EA & Bajah, ST. 1975. *The Namutamba Pilot Project.* ICEE Evaluation Report No 1. Ibadan: ICEE, University of Ibadan

Yoloye, EA (ed). 1978. *Evaluation for Innovation: African Science Education Programme Evaluation Report.* Ibadan: Ibadan University Press

Yoloye, EA & Bajah, ST. 1981. *A Report of 20 Years of Science Education in Africa.* Accra: SEPA

Yoloye, EA. 1994. *Humanism and the Science Curriculum.* Science Teachers' Association of Nigeria (STAN) Position Paper No 5. Ibadan: STAN

World Bank. 1995. *Priorities and Strategies for Education. A World Bank Review.* Washington, DC

# 2

# The role of science and technology in development

*Phineas Makhurane, University of Science and Technology, Zimbabwe, and Michael Kahn, Centre for Education Policy Development, Johannesburg*

## ABSTRACT

The authors begin by presenting a historical perspective on the role of science and technology worldwide, with particular reference to Africa. They address questions such as: Is development linked to social and economic systems? Who defines development? For what kind of development should Africa strive? What kind of science and technology education best promotes this development? What is the relationship between science and technology and development? Do realistic or deterministic views of science and technology better suit development in Africa? The chapter provides evidence to support claims, analyses trends in the role of science and technology in development for past and current practices, and proposes suggestions for Africa in the future.

## DEVELOPMENT: AN AFRICAN PERSPECTIVE

> Technological dependence lies at the heart of all dependencies. Therefore, we in the developing countries should evolve a technological capacity appropriate to our own conditions; select technologies and adapt them to our own economic and social infrastructures in the context of our own culture and way of life.
>
> Dr Rodrigo Borja, President of Ecuador

### The past

Africa is rich and diverse in resources — it has 97 % of the world's chrome, 85 % of its platinum, 50 % of its palm oil and 33 % of its coffee (United Nations Economic Commission for Africa) — terrain and people. Ancient civilizations in Africa such as those of the Nile Valley exploited fertile soils, water and settled labour to develop class structures that could evolve technologies such as the pyramids, wheeled chariots, and the iron mines of Meroe. Similar civilizations produced the bronze technologies and mammoth earthworks of Benin in the west; the clay-domed steel furnaces on the savannah plains of the east; and the iron mines and stoneworks of the kingdoms of Greater Zimbabwe.

European accounts written at the time of early contact marvelled at the equitable social organizations and inventive technologies compared with those at home — a state of mind that did not last long before it became more convenient for Europe to view Africa as backward, and as a source of minerals, agricultural raw materials and slave labour. Profits helped fuel capital formation and industrialization in Europe.

Cheap labour is a necessary element of industrialization. Despite a variety of systems of chiefdoms, kingdoms and empires, in most parts of Africa people controlled their mode of production, the land. This system of subsistence farming has likely been both the curse and the blessing of the continent. By contrast, in Europe, centuries ago land enclosure acts separated peasants from their economic roots to become compliant sources of cheap labour.

The Berlin conference of 1884 formalized the end of the European struggle for spheres of influence in Africa with the artificial creation of the current nation states. Exploitation of Africa's labour and mineral and agricultural resources characterized the relatively short era of imperialism that followed. What little development occurred was to facilitate the exploitation of these resources and to develop tastes for manufactured goods and thus expansion of the market.

The scramble out of Africa was therefore more rapid than the scramble into Africa, following the realization that continuing economic imperialism could be achieved without the expense and inconveniences of formal imperialism.

## The present

On the whole, Africa continues to be a continent of subsistence farmers and pastoralists — some claim that Africa's contribution to the world's industrial output is only 2 % (UNECA). Urban salary scales assume that a wife is at home on the farm feeding the family. Rural areas further subsidize cities through the sale of excess food at subsidized prices and the purchase of manufactured goods. The livelihood of African subsistence-level farmers remains relatively untouched by events in cities. Price controls on the basic food crops provide farmers with little incentive to produce an excess. Thus the smallholder sector is an unsatisfactory base for capital formation. This is generally an urban phenomenon where small elites become disproportionately rich through the exploitation of contacts with government, from the extraction of minerals, and from servicing expatriate communities. Small industrial sectors focus on import substitution, off-shore industries making large profits from processing raw materials. South Africa is a notable exception, where, as in Europe, the masses (in this case black) were systematically driven from the land to become workers for the substantial industrial sector. However, with expanding populations and the concomitant need for more land, environmentally friendly practices such as shifting cultivation become impossible, friable tropical soils deteriorate, and primary forest areas are disappearing at alarming rates. As a result, many African countries can no longer feed themselves.

Everywhere in Africa the notion of the nation state is under attack from within by ethnic minorities dissatisfied with corrupt national leadership, and from without by multinational corporations. Monolithic economies, overdependent on a single foreign exchange earner such as copper in Zambia and petroleum in Nigeria become vulnerable to change in world markets and are collapsing. In many African countries, by almost every measure people are worse off than they were 30 years ago.

Ghana and Uganda are two countries in Africa that are currently experiencing an upward swing of their economies. They were the subjects of a report commissioned by UNECA and prepared by the Foundation for Research Development (FRD), South Africa. Table 2.1(a) shows some human development indicators in the two countries selected from the report. Table 2.1(b) shows some science and technology indicators.

Most of the countries ranked in the bottom 97 least developed countries are in Africa. Africa is also the least developed continent in terms of science and technology, if indicators such as journal articles and citations are used as a guide. There are numerous theories to explain this underdevelopment of African countries.

The environmental thesis argues that the harsh conditions in the North necessitated development of the advanced technologies that enabled it to dominate the world, including Africa. The assassinated Guyanese sociologist, Walter Rodney, in *How Europe Underdeveloped Africa,* suggested a deliberate policy of underdevelopment through mercantilism and market expansion in a continent ravaged by the legacy of slavery — an earlier exploitation of cheap labour that was one of the bases for European capital formation. Pakenham, in *The Scramble for Africa,* comments on the almost total ignorance of Africa by Europe until the end of the last century: '... beyond the trading posts of the coastal fringe, and strategically important colonies in Algeria and South Africa, Europe saw no reason to intervene.' Yet with the impact of a 'romantic nationalism' and economic depression, Africa became '. . . a lottery ticket, and a winning price [that] might earn glittering prizes'. Any theory to explain underdevelopment in Africa must be complex, and forms the background for consideration of the role of science and technology.

Africa was the last continent to modernize and participate in the debate on the role of science and technology in development. However, though a latecomer, discussion and expectations have been intense in Africa, especially during the 1960s and the first wave of decolonialization. Regrettably, as table 2.2 indicates, there is little to show for some 20 years of investment in science and technology, and continuing expectations have become something of a cargo cult. This chapter argues that the issue involves misunderstandings of the very nature of science and technology, of innovations, and of intellectual and economic hegemony.

 

**Table 2.1(a): Rankings of Ghana and Uganda among developing countries, based on human development indicators***

| | Ghana | Uganda |
|---|---|---|
| Human development index (1992) | 62 | 80 |
| Life expectancy (1992) | 58 | 96 |
| Access to safe water (1988–91) | 61 | 95 |
| Infant mortality (1992) | 56 | 74 |
| Daily calorie supply (1988–90) | 79 | 87 |
| Child malnutrition (1990) | 65 | 62 |
| Adult literacy (1992) | 52 | 68 |
| Mean years of schooling (1992) | 44 | 76 |
| Radios (1990) | 26 | 65 |
| Real GDP per capita (PPP$) 1991 | 75 | 70 |
| GNP per capita (US$) 1991 | 63 | 91 |

*Ninety-seven developing countries were ranked to reflect their comparative performance on selected aspects of human development.

**Source:** Human Development Report 1994, UNDP, New York. UNECA, 1995, in FRD.

## SCIENCE AND TECHNOLOGY

> The hand axe from Olduvai, over a million and a half years old, a first thing made by man, prefigures the whole world of making and shaping. No earlier artifact exists on earth, and all art and technology begins here.

This quotation from the brochure of a major exhibition of African art reveals the intimate relationship between art and technology and their basis for the human activity called science.

However, history dictates that societies keep at the cutting edge of technology to avoid domination by others. One may be first through the technological door, but that advantage must be nurtured and maintained. Those who know iron are likely to dominate those who know flint.

There is considerable debate as to the existence of science in Africa, or indeed pre-contact Europe. However, there is general agreement that technology — making — always existed in all societies, including Africa, and preceded science — formal analysis and theorizing — though some argue that technology is merely applied science and see a linear development from science to technology. Others see science as representing objectivity and rationality, whose purpose is to understand nature;

technology as making in ways that help how we conduct our lives. Yet others see both as more messy, human activities that wriggle and fumble as they progress.

The philosopher Fukuyama ascribes a special role to the progress of science in societies as providing an arrow of time to measure development. One of the authors of this article (Makhurane) aligns himself with a rationalist view of science as a culture that may be superimposed on any culture since it is universal, and a culture of hope and undying optimism. This view believes that a scientific interrogation of nature should lead to the same answer, irrespective of who the interrogators are and where they are located. However, as with any discipline or craft, possession of the tools of science is no guarantee of rational behaviour, compassion and humility. Furthermore, the development of science appears to go hand in hand with the development of economic and political hegemony — witness the riches created by contributions made by chemistry to the textile industry in Germany and Britain at the turn of the century.

**Table 2.1(b): Selected science and technology indicators for Ghana and Uganda**

| | **Ghana** | **Uganda** |
|---|---|---|
| Estimated expenditure on R&D, 1993 (US$ millions) | 4,67 % | 2,03 % |
| R&D expenditure as % of GDP (1993) | 0,08 % | 0,06 % |
| R&D expenditure as % of government expenditure (1993) | 0,40 % | 0,29 % |
| S&T publications & (citations) 1981–92 | 765 (2 772) | 423 (1 350) |
| Estimated number of researchers (1994) | 850 | 800 |
| Researchers per 10 000 labour force | 1,59 | 1,03 |
| | | |
| | **(1990)** | **(1994)** |
| Number of students enroled in higher education (year) | 13,316 | 10,492 |
| Percentage of higher education enrolments in S&T fields | 42 % | 59 % |
| | | |
| Balance of payments (US$ millions) in: | **1988** | **1992** |
| • high-technology goods | –100,8 | –114,3 |
| • medium-technology goods | –435,1 | –176,6 |
| • low-technology goods | +395,5 | –77,0 |

**Source:** UNECA, Development of Appropriate Science & Technology Indicators, 1995.

**Table 2.2: Changing indicators for Ghana and Uganda**

| | Ghana | Uganda |
|---|---|---|
| Energy use (oil equiv) per capita (kg) (1971) | 107 | 58 |
| Energy use (oil equiv) per capita (kg) (1991 | 130 | 25 |
| Energy use (oil equiv) per capita (kg) (1993) | 96 | 21 |
| Energy imports as % of merchandise exports (1971) | 8 | 1 |
| Energy imports as % of merchandise exports (1992) | 52 | 73 |
| Annual average change in forest (1970–89) | –0,8 | –0,8 |
| Infant mortality rate (per 1 000) (1982) | 98 | 116 |
| Infant mortality rate (per 1 000) (1993) | 76.2 | 99,2 |
| Education expenditure (as % of GNP) (1960) | 3,8 | 3,2 |
| Education expenditure (as % of GNP) (1990) | 3,3 | 2,9 |
| Health expenditure as % of GNP (1960) | 1,1 | 0,7 |
| Health expenditure as % of GNP (1990) | 1,7 | 1,6 |
| Population per physician ('000) (1970) | 13 | 9 |
| Population per physician ('000) (1990) | 23 | 14 |
| Science publications recorded in ISI database (1985) | 79 | 39 |
| Science publications recorded in ISI database (1994) | 114 | 91 |

**Source:** UNECA, 1995, in FRD.

How we define science and technology becomes problematic when we exclude the social contexts within which they are practised: how we define them colours our expectations of what science and technology can contribute to development.

## DEVELOPMENT

It is insufficient to define development as the improvement of the quality of life and wellbeing of the ordinary citizen. We must develop indicators to refine and quantify the definition.

Anthropological studies reveal that within some traditional tribal structures, by accepting the authority of the chief, one was guaranteed an education for one's children, a job in administration and support in one's old age. All these are commonly used indicators of development. For Fukuyama (1992), without the attainment of liberal democracy together with a free market, development cannot occur. Using these indicators, 19th-century UK and the Soviet Union could be described as 'developed'.

We must consider other indicators to evolve an understanding of development. These would include: (1) the degree of social cohesion; (2) the extent of literacy; (3) the value attached to productive activity; (4) the role and nature of its class formation; (5) the quality of its health and nutrition; (6) equality of access to opportunities; and (7) mortality rates and life expectancies. Table 2.3 shows some development indicators in Ghana and Uganda.

**Table 2.3: Development indicators for Ghana and Uganda**

| | Ghana | Uganda |
|---|---|---|
| Population per doctor (1990) | 23 000 | 14 000 |
| Maternal mortality per 100 000 live births (1992) | 1 000 | 550 |
| Mortality of children under 5 yrs (per 1 000) (1992) | 129 | 205 |
| Population with access to health services % (1985–91) | 60 | 70 |
| Population with access to safe water % (1988–91) | 54 | 15 |
| Population with access to sanitation % (1988–91) | 42 | 31 |
| Mean years of schooling (1992) | 3,5 | 1,1 |
| % of paved roads in good condition (1988) | 28 | 10 |
| Motor vehicles per 100 people (1989–90) | 0,8 | |
| Telephones per 100 people (1990–92) | 0,5 | 0,3 |
| Television sets per 100 people (1990) | 1,5 | 1 |
| Radios per 100 people (1990) | 27 | 11 |
| Paper consumed per capita (kg per 1 000 people, 2 990) | 300 | 0,05 |

**Source:** UNECA, 1995, in FRD.

Whatever indicators are used, ultimately, defining development is a value-laden process done within varying social contexts. For the capitalist, for instance, development is achievement of wealth with a minimum of interference. However, South Africa, a developing country of sub-Saharan Africa with an emerging democracy, has begun to define its indicators differently to bring them in line with national goals stated in the White Paper on Science and Technology in South Africa (1995). The White Paper states that the goal for development is a future '. . . where all South Africans will enjoy an improved and sustainable quality of life, participate in a competitive economy by means of satisfying employment, and share in a democratic culture'.

The data for Ghana and Uganda is indicative of the reality in many parts of Africa that needs no further elaboration. Many countries have a long way to go before achieving even a handful of the stated indicators.

## SCIENCE, TECHNOLOGY, AND EDUCATION IN DEVELOPMENT

### Science and technology

Much scientific research in Africa duplicates or is an extension of western research. Thus, development of Africa's scientific and technological resource base is often measured in terms of journal articles published in the industrialized North and citations to these articles. Doing so can present a biased picture that may result in inappropriate planning and development as well as affecting the flow of foreign investment. Table 2.1(b) (UNECA) shows some science and technology indicators.

Basic research in small particle physics has little application to development in Africa, even were there funds to support it. Without closer analysis, it would seem that so would basic work in, for example, astronomy. However, in comparison with skies in the North relatively little is known about the southern skies. Therefore, Africa would have an advantage over the North in this field, less financial support would be needed than for particle physics, and engagement by African scientists in high science would keep them at the cutting edge with their international colleagues. Though clearly it is inappropriate and impossible financially for African countries to engage in all aspects of high science, certainly there is a need for scientists in Africa to maintain close contacts with high science through journals, meetings, exchange fellowship programmes and so on. To keep African scientists at the cutting edge, there may be a need for both national centres of high science throughout the continent, such as at the University of Cape Town in South Africa, and for regional centres, such as the International Institute for Tropical Agriculture (IITA) in Nigeria. Individuals and donors, rather than cooperative vision of African governments, have been responsible for regional centres of high science. Considering the scarcity of resources and the brain-drain, African countries should consider support to regional centres of excellence that address problems in crop and livestock management, thus directly contributing to development. However, there may be indirect relevance in the theoretical astronomy and mathematics research at Cape Town. Familiarity with any branch of high science may enable more inspired work in directly relevant low science. The fundamental base of high science cannot be prescribed. Provided its research agenda has intellectual integrity, its contribution to national development lies in training skilled researchers whose ability to analyse and operate critically forms an essential component of any society to develop — though one hopes that university training can avoid the '. . . dominance of culturally and socially inappropriate curricula and structures which do not reflect the interests of the major stakeholders' (Gaidzanwa, in UNECA, 1995). One does not expect fundamental breakthroughs from isolated science departments in Africa. However, one does expect them to contribute to the stock of skilled personnel with the skills and respect for rigour needed to contribute to the economy wherever they find themselves.

We define 'high science' as research that pushes back the frontiers of knowledge. 'Low science' — which has more direct relevance to development in African countries — we define as using existing knowledge to solve pressing issues in fields such as health, family planning, agriculture and the production of goods. Low science enables people to add value to their resources, is itself problem-solving in its execution, and blurs distinctions between science, technology and, on occasion, sociology. The knowledge base in low science is thus as important a resource to African countries as their mineral and agricultural resources. The locus for any high science conducted in African countries is clearly the universities or international research centres. The locus for low science would be universities, government laboratories, extension agencies, industry, and consultancy centres in informal sectors, such as those established by the University of Science and Technology, Kumasi, throughout Ghana. South Africa's White Paper on Science and Technology (1995) is exemplary in the ways it uses policy to link high science, low science and technology in pursuit of social and economic goals.

Industry is also a focus for the practice of high science, low science and technology. Where this is found in-country, there are clear advantages to linking academics with industry so that research would be pulled by market demands. Governments in Africa should promote such research through passing appropriate patent laws, implementing credit schemes, and giving tax incentives to individuals and firms that support applied research. When the research capabilities of multinationals are located outside Africa, as is so often the case, there may be a mismatch between these science activities and the demands of Africa.

## Education

Countries in Africa have invested heavily in education. Kenya, for example, spends about 40 % of the recurrent budget on education — well over 50 % if education components of other ministries such as health and agriculture are taken into account. And this does not consider expenditures on schooling by parents who in rural areas may spend 60 % of household cash incomes on educating their children. Yet despite an emphasis on science and technology in the curriculum, economies throughout the continent are in disarray. However, the informal economy, typified in the 'market mammies' of West Africa or the bush mechanics in East Africa, flourishes. On the surface, it appears that education hampers the economy! Poorly or uneducated traders and mechanics outperform educated bureaucrats: legions of school-leavers and university graduates are unable to find gainful employment, unlike their uneducated age-mates in the informal economy.

Perhaps the school curriculum, particularly the science and technology components, is dysfunctional? Perhaps the teaching of them makes these subjects unpopular and mysterious? Schooling is an elite activity and the higher the rank in the educational pyramid, the more elitist it becomes. For example, in Tanzania only about 5 % of students who enter secondary school enrol in senior secondary classes. They are clearly on a career path that will take them to universities and jobs as medical doctors, government functionaries and so on. This high investment scarcely impacts on the broad economy, and the skills developed remain locked up in the public sector, scarcely touching the huge informal sector.

Making a transition from a low-income survival activity in the informal sector to a more complex one is not easy. Governments could divert resources invested in the formal educational system to provide the informal sector with the necessary inputs to facilitate such a transition. Performance-linked incentives provided by government to universities in Ghana redirected research. Universities did so by establishing consultancy services in the heart of the manufacturing and informal sectors. This is a model that may be worth pursuing elsewhere in Africa. So may the experience in Mauritius, where they introduced a multidisciplinary approach to science and technology courses that emphasizes practical training and exposure to the market place. The South African White Paper (1995: 10) summarizes an approach to education and training that other African countries should consider:

> Education and training in an innovative society should not trap people within constraining specialties, but enable them to participate and adopt a problem-solving approach to social and economic issues within and across discipline boundaries . . . Basic inquiry, as opposed to a formula-driven approach, is absolutely essential, particularly at the universities and technikons, which deal with young minds.

The type of relationship between researchers, extension agents and farmers that has worked in agriculture also may be a model that informal science and technology educators would wish to examine. It could provide a wedge to bring more advanced technologies into the informal sector. Perhaps what holds Africa back is a lack of willingness of her peoples to believe in themselves, their ability to change their circumstances and their preparedness to invest in their own futures — though they are the continent's most valuable resource. Becoming an innovative software developer, for example, does not require massive financial investment, as Sri Lankans have so amply demonstrated.

## CONCLUSION

Unlike the Organization for Economic Cooperation and Development (OECD) countries, no country in Africa other than South Africa has invested in the type of policy research needed for what have come to be called foresight studies to determine possible long-term outcomes of science and technology decisions. Instead, planning in African countries has become dominated by the views and actions of donors such as the World Bank and the International Monetary Fund (IMF). Such organizations generally provide the funds for the research that informs planners. Inevitably such work lacks a strong African perspective. Therefore, African countries must themselves undertake to sponsor research into development, and into science and technology indicators. If need be, bodies such as the Economic Commission for Africa (ECA) or the Organization of African Unity (OAU) should sponsor appropriate standardization and training.

## REFERENCES

Foundation for Research Development, South Africa. 1995. *Development of Appropriate Science and Technology Indicators for Africa.* UNECA

Department of Arts, Culture, Science and Technology, South Africa. 1995. White Paper on Science and Technology

# 3

# Curriculum innovations and their impact on the teaching of science and technology

*Mike Savage, African Forum for Children's Literacy in Science and Technology, Nairobi, Kenya*

## ABSTRACT

This chapter describes classroom practice in Kenya during the 1970s as well as how changes in syllabi, curriculum materials, teacher development and examinations supported inquiry science learning. Current innovations in science education are described and then analysed to identify factors that are critical for the dissemination of innovative practice.

> ... problem solving skills can be applied to a wide range of work settings and can enable people to acquire job-specific skills and knowledge in the workplace. (World Bank, 1994:10)

> It occurred to me, then, that of all the virtues related to intellectual functioning, the most passive is the virtue of knowing the right answer. Knowing the right answer requires no decisions, carries no risks, and makes no demands. It is automatic. And it is thoughtless. (Duckworth, 1987)

## THE KENYAN EXPERIENCE

Much of what I describe is of schools in Kenya during the 1970s — a country I know well — and of primary education, my area of interest and experience. I describe Kenyan primary schools in detail to provide texture and a picture of what I regard as good science teaching. During the same period, equally exciting work could be found at all levels in many English-speaking countries throughout sub-Saharan Africa as a result of programmes such as the African Primary Science Programme (APSP), the Science Programme for Africa (SEPA), and Nuffield Secondary School Science. These programmes encouraged educators to engage with their own realities and permit themselves to be inspired and sustained by the creativity of children and the inventiveness of teachers. I have visited science classrooms in over 20 countries on four continents and though the most dismal I have seen have been in Africa, so have the most exciting, often taught by creative but poorly qualified teachers. As the most outstanding of a number of outstanding teachers in Africa once said, 'It's just a matter of struggling.'

### 'It's just a matter of struggling'

Samuel Githinji, the teacher who remarked that 'It's just a matter of struggling,'[1] was one of those teachers whose classrooms one never wanted to leave. Every moment spent there was a joy. One never knew what to expect when visiting Githinji. A permanent feature of the classroom was the science store in a fenced area adjoining the school; a pupil-constructed mud and wattle building with row upon row of shelves burdened with science equipment. Equipment to Githinji and his students meant materials salvaged from the environment, such as bottles, tins, scrap metal, old car parts, wire and lumber. It also meant home-made tools such as spring balances made from inner tubes, thermometers from old ball-point pens, water drop magnifiers and microscopes, weather recording instruments and so on. On one occasion the fenced area had a 30 ft (ie almost 10 m) high windmill that drove a circular saw, every part made by students. Once there was a 5–6 ft (1,5 m–1,8 m) concrete and mud sphere with the globe painted on it. For a few hours I understood longitudes, latitudes and time zones as students patiently explained using shadows of sticks stuck to the massive globe with lumps of clay. I've seen a relief map of Kenya in that fenced-in area, made to scale with Mt Kenya about 4 ft (1,2 m) high; an experimental farm to investigate ways to irrigate crops to conserve water; brick-making kilns and charcoal-making fires to explore more effective technologies. All that before one entered the school building!

At any given moment, half the class were not even in Githinji's classroom, they were working outside. When all 50 gathered together it was always a surprise that there were so many pupils in the class. Groups inside worked with materials, arguing vivaciously with each other, Githinji or the visitors. Children have asked me questions I could not answer. They have asked me to settle arguments. Students once even asked for one of my hairs to see if it was better than theirs for the hygrometer they were designing. No two groups ever seemed to work on the same topic. Some children could be making instruments for the classroom orchestra, perhaps testing different wires to determine their breaking points. Others could be updating accounts of sales and purchases for the classroom farm. I remember one solitary boy experimenting with strips of metal to find which was the best for making a clapper for an electric bell. When I asked him why he was doing that, he explained that Githinji had caught him misbehaving and had ordered him to make one as punishment. I have seen groups replicating Faraday's investigations of a candle and making high-powered slingshots and rubber-powered guns. I have seen groups designing and making toys and playground equipment for younger children in the village and special equipment for the disabled. A local craftsperson, such as a herbalist, could be in class explaining aspects of her trade. On one occasion the class visited a nearby teacher training college where a group of women students had come out with an inexplicable rash. The local newspaper reported that a medical officer had identified college food as being responsible for the outbreak. Githinji's pupils did not believe this so they visited the college to interview students and collect other data. They correctly identified a new soap powder as the cause.

Reference books written by children filled Githinji's classroom rather than text-books. Over the years, this classroom library became a fount of information about the local community: its flora and fauna; soil erosion sites; maps; areas of knowledge of local experts; local job opportunities; and, most important, pupils' analyses of past examination papers. Despite a total lack of cramming, Githinji's pupils always did adequately in the public examination.

It was not enough to visit the school. To have a more complete picture of the impact of Githinji's teaching, one had to visit pupils' homes. Torch batteries and bulbs wired huts so you could switch on a light to see your way to the kerosene lantern. One detail about these circuits that always intrigued me was the torch bulbs mounted inside larger AC light bulbs. One would sit on furniture pupils made, eat food they had grown, and be protected from mosquitoes by burning incense they had concocted. Some households had wheelbarrows with springs to ease their burden over the bumpy paths. Others had hand-powered sawmills, the handles mounted on large flywheels. Many homes had vegetable gardens, poultry, sometimes rabbits, using husbandry techniques children had developed at school. One mother explained that when her son first moved into Githinji's class she thought of removing him since it seemed pupils were just messing around. She soon realized what he was learning was worthwhile, because he spent his time working on home improvements instead of getting into trouble in the village. Better still, her son had started to take school seriously and studied hard in all subjects, not only science. When explaining science to me, her son once said, 'Well you see, with science you never seem to know. For example, that poison we made for the mole rats. When we put it on the ground by the sugar cane it keeps them away. But then who knows, the sugar might suck up the poison and then when we eat it we might get sick. With science, you never really know, it's always a bit of a mess.'

The Commonwealth Association of Science, Technology and Mathematics Educators (CASTME) awarded Githinji a second grade in 1975 — a year when they deemed no entry merited a first. Though Githinji's work was outstanding, I could describe many equally exciting classrooms in Kenya that I visited during that period.

Githinji was a P3 teacher, that is, he had the lowest possible professional qualification. However, he was a remarkably curious man. Everywhere he went he saw questions and materials to collect for investigation at his leisure. As he once said, 'Collecting has always been my nature. Everywhere I travel and see something of interest, I collect it.' His tutor at training college, Alex Berlutti, himself a remarkable science educator, had reinforced Githinji's innate curiosity. Both Berlutti's and Githinji's formal education had been minimal. It had been insufficient to erode their inquiring minds, but sufficient to set them off as autodidacts, sharpened by some knowledge of the ways rather than the facts of science. Somehow, their formal education had strengthened their self-confidence in their ability to solve problems, rather than eroding it.

In most educational institutions in Africa today, learners and practitioners at all levels distance themselves from their realities and instead engage in abstractions. The joy has been removed from learning, educational practice and research that have become largely irrelevant burdens.

## A village museum, science equipment factory and more

> In contrast to the university-based researcher, the organizer ... gradually becomes recognized by community members as having a commitment to their well-being. The organizer immerses him or herself in the life of the community, learning its strengths, resources, concerns, and ways of conducting business. The organizer does not have a comprehensive, detailed plan for remedying a perceived problem, but takes an evolutionary view of his or her own role in the construction of the solution ... The form they will take is not always known in advance. It is the organizer's task to help community members air their opinions, question one another, and then build consensus — a process that usually takes a great deal of time to complete. (Moses et al, 1989, in St John)

Leonard Kimani was the tutor at the Limuru District Advisory Teachers' Centre. As an organizer, his strategy was simple. He assumed that all teachers had interests that frequently had little to do with their professional qualifications. Kimani identified these interests and exploited them. Innovative work in schools originated from such interests, encouraged by Kimani. The teachers then interested children.

Children in a local primary school had organized a museum in a disused, wooden classroom block with the internal partitions removed. Along the back were rows of cages with chickens and rabbits, each with careful records of different feeds and growth rates. Children failed to convince me that their data proved that feeding rabbits a particular weed prevented them from conceiving, and that when the weed was removed from their diet, they had bigger litters than normal.

The museum had a bones section. I remember a set of femurs, ranging from that of a giraffe to that of a minute shrew. There was a skeleton of perhaps a cat, with an invitation to try to piece the bones together. The table had a saw and various household chemicals tempting one to investigate bone structure and composition.

Torch cells and bulbs littered a table, together with wires, different substances that were conductors or insulators, wire coiled-nails, pupil-made motors, magnetized needles stuck in pieces of cork. All had intriguing questions that provoked investigation.

Mystery substances and liquids on one table demanded to be identified. Another area invited the viewer to build towers and bridges from grass stems and thorns and to test their strength. One important aspect of the museum was the cardboard boxes under each table. Children explained that these were filled with materials to be loaned to teachers when they taught the relevant topics. Without such encouragement, they claimed, teachers usually only used chalk and talk which was boring.

Another school in the same district had a science equipment production factory, run by the children. From so-called junk, the children made science kits for other schools. The kits contained tools rather than demonstration equipment designed for any single purpose. Each kit contained hand tools such as hammers made from large bolts, saws from tree branches with hacksaw blades, and screwdrivers and chisels from tempered six-inch nails. It also contained scientific tools such as magnifiers made from electric bulbs and packing crates with holes for water drops; pegboard, rubber-tubing, and pan balances. There were weather-recording instruments and other useful tools to extend children's ability to investigate.

A third school in the district became a soil-conservation centre. Pupils mapped potentially friable sites and ran a school nursery that grew multipurpose, ornamental and fruit seedlings for sale to parents. Yet another school became the district centre for the analysis of past examinations and preparation of mock papers. As candidates from the district improved their performance in examinations, parents increased their interest and support, both moral and financial, for work done by these schools.

## A need for systemic change

In Africa today there is discussion and despair about the impossibility of changing science teaching for the better. Obstacle after obstacle to implementation is identified, and action recommendations are made in the safety of international conferences that participants suspect will not be implemented. A critical analysis of the Kenya experience provides evidence that our reflections and recommendations may not be too unrealistic and that there are grounds for optimism. The analysis is made using frameworks recommended by participants at ASTE '95 as being those that should underlie effective change models. Governments and donors would be well served by a more comprehensive and authoritative review of past efforts to help maximize future inputs and develop policies that promote quality science education.

### *Change should be incremental, participatory and focused on human development*

Together, American technical assistance staff and their Kenyan colleagues initiated change during the mid-1960s through participation in the APSP. Though there was no fixed time frame, all knew the process would take decades. For the first two or three years, staff at the curriculum centre worked intensively for 2–3 days each week with teachers such as Githinji and Kimani in a few schools, further developing teaching units inspired by regional APSP workshops. Two rural subcentres were established in an attempt to ensure that materials reflected classroom realities representative of different parts of the country. Units were inquiry-based and used materials found in the school environment. Early during this period the team realized there was a need for more experienced staff and selected individuals were sent to the University of Sierra Leone, which at the time had an outstanding inquiry-based degree programme.

By the early 1970s some 40 teaching units and a dozen supplementary booklets for children had been developed and published as a result of this cooperation of pupils, teachers and curriculum staff. As important, a core of skilled, dedicated teachers and educators had evolved with a shared vision and experience. An external evaluation conducted by Yoloye and Duckworth showed that the approach was effective in achieving its objectives when handled by teachers associated with the programme. However, experience showed that teachers not directly involved had difficulties using the units.

In 1972, curriculum staff together with the experienced core teachers embarked on the development of a set of teachers' guidelines to help other teachers better use the inquiry approach and units. The team anticipated implementation problems and deliberately developed these guidelines with teams of teachers from every district in Kenya, training college tutors, teacher centre staff (expanded to 40 from the initial two subcentres), and inspectors, together with a few scientists. These teams, joined on occasion by senior officials including the director of education, visited project schools throughout the year and each December met in different parts of the country to revise materials tested during the previous year and draft the guidelines for classroom trial during the subsequent year. Project materials reflected classroom problems identified by teachers, implementation problems identified by inspectors, and teacher education problems identified by teacher educators. By the mid-1970s the Kenya Primary Science Project knew it was involved in systemic change.

### *Change must be systemic, reflect classroom realities and be sustainable*

As a result of inspectorate involvement in the curriculum development process, teacher development was identified as a limiting factor. Subject inspectors successfully lobbied for funds for large-scale in-servicing. For three years, five-and-a-half-day courses were run in every province in the country for groups of up to 500. Teams of people from the different institutions involved in the development of units and guidelines worked with these district and zonal teams that represented a similar spread of institutions. The goals of these workshops were not to train participants in the use of course materials. Instead the focus was on professional development. Workshop participants were exposed to working with materials, with children, and with other teachers not at the workshop, for whom they ran a one-day in-service course drawing on the resource team, guidelines and units. Zone and district teams spent the last day developing action plans for the following year, creating support structures for local groups of teachers, and negotiating assistance rather than supervision from headquarters. Multimedia resource material for in-servicing teachers were developed using the experience of these efforts as well as radio and television broadcasts and newspaper releases to bring the changes to the attention of the public.

As a result of the involvement of training college tutors a demand grew for changes in pre-service teacher development programmes. For several years science tutors representing all 18 colleges worked to develop appropriate print, film, slide/tape and audio material for tutors and students to prepare teachers more effectively for the changes being implemented in schools. From wherever participants were located in the educational system, they identified equipment as another important limiting factor, even though the approach advocated exploring phenomena found in the immediate school environment. In the mid-1960s, with help from the Ford Foundation, a science unit had been established at the curriculum centre to address this problem. Initially the unit designed prototype apparatus with a view to mass production and used specially designed mobile science workshops to tour colleges and train students in their use. This was modified in the light of experience and instead workshop technicians became part of the curriculum development and implementation teams, exposing participants to skills of using local materials to make tools that facilitate learners' inquiry. Thus, though unstated, a deliberate policy decision was made to create an environment that supported and encouraged maximum use of local resources rather than a dependency on centralized equipment production that could never fulfil expectations.

#### *Curriculum goals, materials, teacher support services, syllabuses and exams must not be in contradiction*

Syllabuses were identified as yet another factor limiting change. Since all parties had been involved with and were supporters of the new approach, changing the syllabuses met with little resistance. In 1976, the primary science syllabus became a slender document stated solely in terms of inquiry skills expected at each grade level through investigation of locally available phenomena. The primary teacher training syllabus became similarly couched in terms to promote inquiry into phenomena, learning and skills to provide an enabling learning environment.

A striking example of the effectiveness of the participatory change model is the way examinations changed in Kenya during the mid-1970s. Examinations are repeatedly identified as major constraints on curriculum change. Fix examinations, the argument goes, and everything else follows. The only reason I can identify for their continued tyranny is that the secrecy surrounding them precludes participation. There is little that any individual or organization can do other than strive for the good performance of their child, school or village. They cannot do anything about the nature of the examination nor can they do anything to change it.

Kenya changed primary leaving examinations so that they assessed inquiry-based learning and understanding rather than memorization. Kenya did so as a result of consistent pressure from the growing team of teachers, teacher educators, curriculum staff and examinations council researchers that participated in an increasingly coordinated effort to introduce inquiry science learning. Together they identified examinations as a major limiting factor and began systematically to develop and field test items better suited to their vision of what science learning should be. In the face of evidence and reasoned argument, examinations gradually changed. An important component of the acceptability of the change was the regular feedback given teachers on student performance through a newsletter sent to schools.

The primary science paper now has a moderately fixed structure of which everyone is aware. Some 20–25 % of the paper still tests memorization, but of knowledge deemed essential or likely to have been gained from close observation of nature — an aim of the syllabus. Such knowledge is in the realm of disciplines such as health, agriculture and environmental studies. Another domain of knowledge tested is that most likely to have been gained from inquiry or experiment into common phenomena. Inquiry skills such as the ability to transform data from one form to another, to look for patterns and interpret data, to predict and design experiments, and to evaluate conclusions form over 70 % of the examination paper in Kenya. The most difficult domain to examine in paper-and-pencil multiple-choice tests are practical skills of inquiry such as observation, experimentation and measurement. Yet the primary school-leaving examination paper in Kenya always has some items that do so. Since Kenya is a multicultural society, the examination paid close attention to the cultural context within which items were set, including the culture of girls. Item analysis showed this to be important in determining the performance of different groups of people and, with the exception of elite schools, rural children now generally outperform those from urban areas, and girls perform equally well on most item types. For example, children from nomadic societies performed best on an item asking pupils to decide how the new moon looks in Kenya — the skill of observation of nature being an objective of the syllabus, and the night sky being equally accessible to every child. They were followed by children from coastal areas where many families are Islamic, then by children from farming communities. Children in elite schools chose the option showing the new moon as depicted in their imported textbooks! The protest from elite families that this item was biased is an indication that there is no such thing as a culture-free test; that tests which make such claims in reality favour elites; but, more important, that a participatory approach is essential even in setting examinations in order to maintain a power balance between different members of society.

### *A need for critical mass*

After 15 years of slowly introducing inquiry science to Kenya's primary schools, some thought that critical mass had been reached. Inquiry science had such an extensive network of supporters that one thought it had become embedded in the system. We were wrong. A politically motivated rapid expansion of the educational system together with its restructuring and pressure from large donors almost overnight caused inquiry science to disappear. The educational system of Kenya was reformed and reform means change, not evolution, regardless of the quality of what already exists. Ill-advised policy decisions as much as economic, professional or cultural realities were responsible for the deterioration.

My analysis of the Kenyan experience requires a postscript. The situation has deteriorated to the extent that the state has openly accepted its inability to deliver and has accepted a need for sharing costs with consumers. Yet there are positive aspects to this trend. Local communities and school clusters increasingly make decisions on issues such as what books to buy and raise funds to hire consultants to run professional upgrading courses for teachers. Large industries and manufacturers are realizing that supporting schools with supplementary materials not only demonstrates their willingness to contribute to nation building, but also improves their corporate public image and is possibly a more effective way to reach potential consumers than conventional advertising techniques. Newspapers increase their circulation by including supplements designed for the young. Unwittingly perhaps, the state has unleashed other sources of support to education and perhaps in the future may make policy and curriculum changes that further decentralize the process. Current pessimism may be unwarranted, our timescale too short, and our vision inadequate.

## CURRENT MODELS OF SCHOOL CHANGE

Little of what I have described can be seen in Kenya today, or in most African countries. Rapid expansion of educational opportunities, increasingly overcrowded syllabuses, a terrifying deterioration of economies, and political destabilization are some of the causes. However, the past few years have seen a re-emergence of innovation in science teaching in Africa and there is sufficient evidence for renewed but cautious optimism. I am delighted to be able to provide more recent examples of exciting science teaching practices.

### The Zanzibar science camps

A senior science educator I know spends weeks at workshops exploring phenomena, developing teaching units, and being argued down by primary schoolchildren and teachers. His professional self-confidence is such that he welcomes such opportunities. He is sufficiently confident as a science educator not to feel threatened by his ignorance. It is worth mentioning that this science educator is the Honourable Omar R Mapuri, Minister of Education, Zanzibar.

The venue for the Honourable Mapuri's enthusiastic involvement is the Zanzibar Science Camps. Every year during the December school break, many of the island's scientists, science educators and teachers congregate at Nkrumah College on the shores of the Indian Ocean. They run a three-week residential science camp for Form I students, and sometimes upper primary pupils. Zanzibar has organized such camps for seven years.

The camps project would be unthinkable without the participation of two exceptional people. One is Professor Mohammed Bilal, formerly dean of Sciences at the University of Dar-es-Salaam, then principal secretary, Science, Technology and Higher Education and currently chief minister of Zanzibar. The other is Professor Bob Lange, Brandeis University, Waltham, USA. Their vision, energy and dedication led the project through the initial years until that vision and energy was transferred to the educators of Zanzibar.

 

The camps continually evolve. As participants identify problems, solutions are developed. An early realization was that there were not enough girl campers and that teachers from schools which had sent students should also attend the camp. During the third year, organizers became concerned that much of the activity conducted with students was simply more of the traditional teaching usually done in schools. As a result they invited two African educators based in the United States, Hubert and Becky Dyasi from the Workshop Center, City College, New York. The project ignited with enthusiasm as students became involved in inquiry rather than traditional learning. The more the camps developed along these lines, the more the word spread through Zanzibar that something special and exciting was taking place. Visitors began attending for longer and longer periods. First more junior members of the Ministry of Education came to work, such as curriculum, inspectorate and examinations staff; then more senior staff, such as the Chief Inspector of Schools and the Planning Officer. In the fifth year the Principal Secretary spent every moment he could spare working at the camp with materials, students and teachers. In the sixth, the Minister himself attended, not as a guest to open and close the session, but as a participant working side by side with students and teachers.

It is difficult for me to select from the multitude of exciting incidents I have witnessed at the Zanzibar camps. I remember the look of astonished pleasure on a girl's face when, on an excursion to a mangrove swamp, she said, 'You mean we have to learn the language of trees?' I recall the surprised pleasure of the director of the Institute of Marine Biology at the students' sophisticated discussion of what they had seen. This lively session was with the second group of students that visited the swamp. Discussions with the first group had been stilted and forced, the director himself doing most of the talking. Reviewing that session, the resource team decided to divide camper students into groups and ask them to discuss the visit of the previous day. Their discussion lasted an hour and a half and could have gone on much longer. Resource team members joined student groups to enter into debates. During the class wrap-up session, instead of the director doing most of the talking, students repeatedly interrupted to amplify his comments, ask questions and argue with him.

Once in frustration, I half-jokingly threatened to kill Raschid Scheiff, an 'A' level teacher at Fidel Castro Secondary School on Pemba Island, if he did not write up a series of lessons that I had observed him teach. A group of camper students started work immediately and continued for several hours after a brief introduction by Raschid. 'Try to find out as much as you can about the liquids on your desks. Use anything you like in the laboratory and if you need anything else I shall try to find it for you.' As they tried to identify the liquids, all easily found in Zanzibar, students smelled, formed drops and filled containers to the point of overflowing. They used strips of newspaper to separate colours, and juice from hibiscus flowers as indicator. They made layers of liquids, watched seeds and other object sink or float on the liquids, raced drops down sloping surfaces and filled small jars with the different liquids. They also discussed what they were doing and why. They argued about what constitutes a fair test and recorded their results with increasing sophistication. At Raschid's suggestion, after completing each investigation, groups wrote their results on the chalkboard and scrutinized them for patterns. Students interrogated groups whose results were inconsistent with the rest of the class and, if necessary, asked them to repeat their experiments in a standardized way. Students began to order the liquids and notice clusters of experiments that gave similar ordering. Without using technical terms, they began to use concepts relating to surface tension, viscosity, density and pH to explain their observations. When they appeared to have a good understanding of the phenomena, Raschid gently introduced the scientific terminology. In short, the series of lessons was as effective a unit as I have observed on the physical and chemical properties of liquids.

I recollect a group of resource staff exploring seeds before teaching the topic to camper students. They established that some seeds sink in water and others float. They went on to investigate seed behaviour in other liquids. As they heated seeds in water, they noticed that some slowly rose to the surface, paused for a few moments, then sank again. The process repeated itself over and over. The group investigated this dance of the seeds for an extended period before explaining it to their satisfaction.

I believe this experience of making sense of the world is what led to the growing excitement of participants at the Zanzibar camps. Any encounter with phenomena rapidly leads to puzzlement, whether we are primary pupils or university lecturers, and understanding is layered. Our active extension of understanding is exciting and such experiences lead to feelings of confidence, self-empowerment and a knowledge that one, rather than external factors, is in control of one's learning. Not all participants at the Zanzibar camps lost their fear of exposure as rapidly as Raschid or the Honourable Mapuri; not all realized so quickly that their painfully acquired knowledge enabled them to be better inquirers and teachers. However, as participants increasingly experienced inquiry, their investigations became more authentic. With a realization that they were not being asked to discard their knowledge but to use it to expand their understanding, they became less anxious and they too began to experience an excitement that became contagious. Participants began to reflect about what was happening to them and to see ways that they could use their knowledge to promote children's inquiry rather than to teach facts. As they inquired, students' excitement fed back into an ever-growing loop.

The identification of problems during camp sessions affected other aspects of the education system. The ministry has established a science teaching centre at Nkrumah College, well stocked with computers, desktop-publishing equipment and books. Basic science materials have been sent to all secondary schools. The ministry has set up teacher cluster groups and relieved selected teachers of part of their teaching loads to organize workshops and interschool visits. Everybody in the system has become engaged in curriculum development. At workshops, groups develop resource booklets that are accounts of their explorations of materials and of the responses of students, rather than prescriptions for teachers to follow. Primary school teachers, lower secondary teachers, 'A' level teachers, teacher educators, curriculum staff, the inspectorate, examinations' officers and so on use these booklets and add their experiences. Resource materials for pre-service teacher education have been drafted for trial in the training college. Examinations staff are beginning to develop items better suited to the inquiry approach. A group of women spearheaded by the Planning Officer and Chief Inspector of schools is developing ways to work within the community to advocate greater participation by girls in science and mathematics. Training video tapes have been made. The ministry has allocated three full-time staff and a vehicle to help coordinate ongoing project activities. Students organize village and national level science fairs each year. Donor projects concerned with education and environmental protection, such as ODA, DANIDA and the World Bank, have a nationally driven framework within which to work, rather than having to impose new structures.

The project also has an influence outside Zanzibar. For several years Zanzibar has invited delegations from Eritrea and Mbeya, one of the regions on mainland Tanzania. The group of teachers from Mbeya has since started its own curriculum renewal project. The United Nations Educational, Scientific and Cultural Organization (UNESCO) asked Zanzibar to organize a workshop for other countries and island states in the region. In a letter to Ken Prewitt, executive vice-president of the Rockefeller Foundation, Prem Naidoo, a science educator at the University of Durban-Westville, said:

> The Zanzibar Science Camp and its participants from all levels of education, from the Minister of Education to pupils from schools, actively promote the improvement of science education at the primary, secondary and teacher education levels. This is the most innovative project in participatory curriculum development in science education, at a national level, that I have seen in operation. I would rate this programme as cutting-edge and one from which other countries, both developing and developed, have a lot to gain.

In my experience, such deep-rooted curriculum change must be holistic; no magic bullet such as a textbook, interactive radio project or science kit has ever changed a school system. To be effective, many people and institutions in the system must be deeply involved in and committed to change. These and the following factors have been significant in the Zanzibar experience:

- Rooted change is slow. Through the African Forum for Children's Literacy in Science and Technology (AFCLIST), the Rockefeller Foundation and other donors have given Zanzibar time by funding the project since 1988.
- The vision, dedication and support of Dr Mohammed Bilal and Professor Bob Lange during the camp's early years was critical, and as important was their judgement on when to relinquish their role. Zanzibaris now run the project and feel a strong sense of ownership. People are always more strongly committed to implementing their own objectives than those of outsiders.
- The project began in the nonthreatening environment of a camp whose sole purpose was to entertain children. There were no ponderous project objectives or

expectations. Participants were free to evolve their own and could develop solutions without considering the so-called constraints of the education system. An environment that nurtures creativity is necessary for the evolution of creative solutions. People were encouraged simply to have fun, using science as the vehicle. Having fun is addictive, and camp participants gradually worked within the system to maximize such enjoyment.

- Camps had a mix of scientists to keep the science authentic, educators to keep teaching innovative, and children to keep everybody honest. The reference group for curriculum innovators must be the beneficiaries, not their professional peers.
- Camps have the luxury of identifying their own problems rather than having outsiders do so for them. Technical assistance must work for national objectives rather than nationals complying with those of technical assistance personnel.
- During camps, the resource team's roles became blurred; they acted as skilled individuals rather than as ministers, principals, secretaries, or school inspectors and found that they were more innovative than they thought. As many people as possible must participate in curriculum change.

## Linking community with school science in Malawi

In a letter to the technical adviser of AFCLIST, Harold Gonthi of the Malawi Institute for Education reports:

> I am working with the school at walking distance. When children are provided opportunities to be involved they are great achievers. Their own teachers were amazed at the work based on mosquitoes. Several teachers joined the children in their mosquito lesson. It was just great to see these kids at work. A colleague of mine at the Institute remarked after seeing the children's work, 'We don't have bad scientists but we teach them badly.'
> It is exciting work.

This project, supported by the Rockefeller Foundation at the recommendation of AFCLIST, as well as by other donors, is interesting in several ways. The project team assumes that primary school children share the scientific and technological knowledge of the communities within which they live and use this as a basis for learning. The project team increasingly finds it necessary to learn from children by working in primary schools although the project's focus is to develop multimedia materials for pre-service teacher education. One video tape made by the project shows village craftspersons such as the potter and brewer being interviewed about their understanding of the science and technology underpinning their trades. Others in the series show primary school children being taught topics from the science syllabus assuming that their conceptual structures are based on this village knowledge. Tutors' resource booklets accompany each video tape, including one on assessment. The project team that produced these materials includes representatives from the university, inspectorate, the planning, examinations and curriculum sections of the ministry, as well as college tutors. Thus, the project has strong ministry support, easing potential bottlenecks, and provides a model for the current round of curriculum reform, COPE (Community Orientated Primary Education).

However, as the project has developed, so has the realization that participation in the production of materials has been a most important aspect of professional growth. In the course of a second phase, therefore, the project will involve many more college tutors as members of the team developing further video tapes and print materials. These will include materials for agriculture, home science and mathematics, since college tutors in those subjects have clamoured for similar resources. Furthermore, having used the materials, participating tutors have identified that not only students but they themselves have limited experience of investigating phenomena or of teaching inquiry science. Exploration of materials and classroom action research will feature more strongly during this second phase of materials production. During its evolution, project participants have concluded that phenomena and children are the most authentic reference points to judge the effectiveness of curriculum approaches and materials.

## Minds Across, Uganda

Minds Across capitalizes on the chronic shortage of textbooks in Ugandan schools and the curiosity of children about the world around them, to challenge them to write their own. Newspapers and displays line school corridors and classroom walls. Shelves are stacked with booklets authored by children. The four schools that participate in this project have become community libraries for out-of-school youth and adults. The range of titles is extraordinary.

The innovators of Minds Across have harnessed the one resource available in any school, anywhere: the imagination of children. Younger children develop observational skills by drawing and describing familiar experiences. Older children conduct research through inquiry into local phenomena, and gather information from community members, museums and libraries. Children of all ages learn how to plan an investigation and present their ideas. In the words of Katherine Namuddu, one-time project coordinator, 'The problem of illiteracy is more than not just being able to read and write. It's really the inability to communicate what you know.' Minds Across empowers children to value and develop their own ideas. And it empowers teachers to listen to children.

## The Science Curriculum Initiative in South Africa (SCISA)

SCISA provides a model for teacher education. A number of donors support SCISA, including the Rockefeller Foundation, at the recommendation of AFCLIST. It is a network of individuals and organizations concerned with change in science education. They strongly believe that teachers have a key role to play in making curriculum decisions. In a draft national science syllabus, SCISA was the one organization identified as a model for capacity building through the participatory involvement of teachers, parents and students in decision making (Keogh & Salaman, 1994).

SCISA promotes an inquiry approach to science learning and is sensitive to issues of gender and race. Through a network of writers' circles, teachers, educators and others draft responses to policy changes, syllabuses and other documents as well as writing their own curriculum materials. The goal is the empowering of teachers through participation in the process of change.

This innovative teacher education model has strengthened links between the university faculty of education and primary and secondary schools, and in the process altered conventional power relationships. With faculty and SCISA support, during attachments, education students work with mentor–teachers in the school community to identify a science education problem, then they apply action research methods to solve it. The school community and classroom become the locus for research, providing the challenges and site to identify evidence for judging validity. Mentor–teachers become as important researchers as the university faculty. The university recognizes the mentor–teachers' role through accreditation, and expects them to teach university courses and co-author journal articles. SCISA expects change to be incremental and evolutionary. They recognize that participation is a lengthy process, that there are no quick fixes, yet it is a process to which all are committed. Unlike other countries on the continent, South Africa has an economic base and a private sector that can support such a long-term viewpoint. Furthermore, South Africa's recent history has provided overwhelming evidence of the efficacy of a participatory approach to change. Perhaps we should view change as an exponential rather than a linear process.

## The media as a source for promoting inquiry

Schools are not the only source for promoting inquiry learning. Though not as pervasive as in industrial countries, the mass media can be used in Africa to motivate and support children's inquiry, as several AFCLIST-assisted projects have demonstrated. In Sierra Leone, the Home Economics Association taps into traditional communications media channels through youth clubs established throughout the country. They facilitate the youth's engagement in activities such as community health campaigns and income-generating activities. A project of the Wildlife Society of Malawi inspires the youth to use theatre and wall art to galvanize village debate of more rational use of local resources. Spider's Place, produced by Handspring Trust for Puppetry in Education, uses television, radio, comics, and audio and video tapes to reach historically deprived communities in South Africa. Spider's leadership of her gang and their scientific ingenuity repeatedly get them out of the scrapes into which their adventures lead them. The Kagera Writers' and Publishers' Cooperative Society (KWPCS), Tanzania, publishes fifteen thousand copies of a monthly newspaper. It hopes to carry a supplement targeted at the youth. Farmers' cooperatives that have infrastructures such as transport and farmers' centres in most villages distribute the newspaper — an innovative way to have print reach African villages. KWPCS uses the same system to distribute supplementary booklets for children. *Action Magazine*, a heavily illustrated environmental publication, mails 10 free copies to every primary school, secondary school and teachers' college in Botswana, Zambia and Zimbabwe. The Paper Making Education Trust (PAMET) in Malawi helps primary schoolchildren make their own exercise books from recycled paper. Science teachers' associations in Lesotho and Malawi publish students' magazines that promote inquiry science.

All media projects described depend on donors who are concerned with sustainability. Yet all were established specifically to help the poor. As PK Moyo, principal of Matshakayile Primary School in Zimbabwe, explains, 'When it comes, the government grant is very, very little. It can only buy exercise books and very, very, few resource books' (Shankerdass, 1994). Sometimes products such as booklets written by children in Minds Across and copies of *Action Magazine* are the only learning resources to be found in classrooms in these countries.

## ONLY SYSTEMIC CHANGE ENABLES TEACHERS TO CHANGE

Mark St John estimates that the United States spends much less than 1 % of operating education budgets on efforts to achieve change. Any industry that spent such small percentages on research and development could not survive. The percentage Africa spends on change is less, since salaries absorb so much of educational budgets, a situation aggravated by governments' relinquishing innovation to donors. In Africa the situation is unlikely to change in the immediate future. Despite spending over 40 % of the national recurrent budget on education — over 50 % if education components of other ministries such as health are considered — and having introduced cost sharing, Kenya is failing to maintain its educational system, much less change it. That the World Bank is seriously discussing assistance to private education in Kenya is a recognition of decades of failure by government and donors.

It is widely acknowledged that, ultimately, it is teachers who sustain classroom change. It is also widely perceived that teachers are the problem. Design teacher-proof materials, it is assumed, conduct massive in-service programmes to top up their knowledge and the problem will be solved. It's as simple, direct and wrong a solution as experience has repeatedly demonstrated.[2]

Teachers can only change in environments that permit change, and the environment of schools is a complexity of many interrelated factors that has considerable momentum. Yet governments, funders, educational planners and the like expect changing one of the many factors to lead to some magical domino effect — and expect change to be cheap. As any scientist knows, small perturbations rapidly damp down in a system of any complexity. Governments, funders and so on also frequently send conflicting messages to teachers. They expect dedicated service yet pay teachers so little that they are forced to seek other ways to supplement their incomes. Governments expect schools to give children income-earning skills when neither markets nor job opportunities exist. They expect teachers to use child-centred methods of learning then add content to already overcrowded syllabuses without removing anything. They expect teachers to promote thinking skills yet set examinations that test only rote memory. Governments expect teachers to make countless, instant decisions to help children yet do not consult them on major policy decisions affecting classrooms. The litany is endless. Is it any wonder that schools have not changed? Yet many teachers continue to work for change. That truly takes faith: faith and a vision of science education as inquiry, and of change as a participatory process. When I talk of curriculum change, the change I clearly hope for is the promotion of inquiry science learning. There are reasons why I think this should be our goal.

Roberts (1982) identifies seven major curriculum emphases in science education and argues that the greater the range of emphases, the more defensible any curriculum is. The curriculum emphases Roberts (1982: 246) identifies are: 'Everyday Coping; Structure of Science; Science, Technology and Decisions; Scientific Skill Development; Correct Explanation; Self as Explainer; and Solid Foundation.' Inquiry science learning has elements of all emphases except those of correct explanation and solid foundation.

## Science as inquiry, the structure of science, and scientific skill development

David Hawkins defines scientific literacy as what a person deeply versed in some field of inquiry can take to the learning of another. One attribute, he claims, 'is some grasp of the scale-dependence of all natural phenomena, living or inert, from the minute to the grand. The other, that can also be learned from different sciences, is the grasp of the concept, and the art, of successive approximation. This art and understanding are a bond of unity among all the sciences, marking both the solidity of their achievements and their openness to revision. Its failure, by contrast, is prevalent among the fashionable detractors of scientific knowledge and endeavor' (Hawkins, 1994). Rather than being rigid, scientific knowledge can be characterized as being scientific in that it can be modified.

Such understanding can only come through extended inquiry into a few selected phenomena rather than a rapid review and coverage of many topics. Africa needs tinkering, problem-solving citizens, able to judge the appropriateness of evidence. She needs what the (United States) Panel on School Science calls self-governing citizens with '... autonomous intelligence, disciplined to seek and face the truth, and capable of the independent judgment that stands up to wishful thinking and to arbitrary external authority'. Africa also needs highly motivated and trained scientists and technologists to solve our myriad development problems. There comes a stage when exposure to the same educational experiences fails to meet the needs of both groups. Students interested in pursuing science further need different programmes.

Inquiry lies at the heart of science.[3] Scientific process skills are the tools of inquiry; the conceptual frameworks and information its products. Disciplines such as physics, agriculture, technology, environmental studies or whatever would not exist if women and men lacked inquiring minds. Debates such as whether science precedes technology or technology science assume less importance when one recognizes the paramouncy of inquiry. So too do discussions on integrated or environmental education, design education or education that stresses societal aspects of science. I think that school programmes, whatever they are called, that do not give learners an authentic experience of inquiry should not be called science.

Throughout the world, much of the science curriculum reform of the 1960s was inquiry based, emphasizing the structure of science. Projects of that era have been criticized as elitist, concerned solely with producing future scientists. In a sense they were, especially secondary school projects such as the Physical Sciences Study Committee (PSSC), the Biological Sciences Curriculum Study (BSCS) and Nuffield. One must remember, however, that PSSC is an acronym for the Physical Sciences Study Committee and that the originators originally hoped for a joint chemistry/physics course. It was the school system, not the course designers, that arranged for only 20 % of the school population in the United States to study physics. Projects such as PSSC were elitist in the sense that they did their best to provide students with an authentic experience of engaging in science; the element of technology was largely instrumentation to enable further inquiry. They were tough but unquestionably science. I have never heard primary science projects of that era, such as the Elementary Science Study (ESS), Nuffield Junior Science, the African Primary Science Programme (APSP) and 5–12, accused of being elitist. Perhaps that is because children were encouraged to inquire into a wider range of phenomena in primary than in secondary schools. Children's interests and questions took them into technology, societal issues and so on. But the basis remained inquiry, the range of phenomena explored was wide, and there was little criticism from scientists that these projects did not reflect science. On the contrary, the late Jerrold R Zacharias, who was the one individual most responsible for the worldwide upsurge of science education reform of that era, said of APSP, 'I believe of all the science curriculum projects I know about, the African Primary Science [Programme] is by far the best' (Goldstein, 1983).

Lapp has summarized science curriculum innovations in Africa and the United States. Prior to APSP, science teaching in African primary schools was described by Yoloye and Bajah (1981) as '. . . the development of clean and healthy habits, an understanding of nature (plants and animals) and some elementary facts of science'. Teaching methods were dogmatic, based on '... the authority of the teacher and the memory of the pupils' (UNESCO, 1982). Attempts to introduce simple farming principles and techniques met resistance from pupils and parents. The 1961 Conference of Ministers of Education in Africa stressed that schooling must be brought into line with existing African conditions, and set the climate for programmes such as APSP to demonstrate that, given a supportive environment, effective innovation was possible at all levels. Lapp reviews work by Lockard (1972, 1975, 1977), Maybury (1975), Baez (1976), Sabar (1979), Yoloye and Bajah (1981) and others that has evaluated change in science education in Africa. Little has not been tried, and much is picked up again decade after decade. Science curriculum units, teachers' centres, equipment production units have blossomed and decayed throughout the continent. Secondary schools have seen Nuffield-type courses, integrated science and environmental science. Africa has used radio, television, audio and video tapes in an attempt to bring better science education to students. Most science courses have salutary statements of aims and objectives. Yet methods of science teaching in Africa continue to be based on the authority of the teacher and the memory of the pupils as the supportive environments of innovative projects face economic and cultural realities during wide-scale implementation. Only human development endures.

I believe the real issue is not about science or technology. Neither is it about to what extent environmental or societal issues should be taught under the rubric of science. The key issues are what constitutes public scientific literacy; at what stage of schooling should the range of phenomena under inquiry become restricted so that underlying conceptual frameworks clearly emerge for the future scientist;[4] but above all, what changes can realistically be expected in our current socio/political/economic climate?

## Inquiry as children's learning style

Sitting in serried ranks, chanting memorized phrases, is part of the culture of few societies. For African educators to reject inquiry learning as being against the continent's tradition is to forget our tradition — or to remember only the recent, alien tradition of formal schooling. From an early age children everywhere, and more so in Africa, learn by observing the behaviour of adults, by questioning, and by exploring their immediate environment. They learn more specialized skills later under an older, experienced mentor.

There is much current debate on the relevance of constructivism and alternative world views on classroom practice (Jegede, 1995). In an inquiry learning environment children bring their mental and cultural constructs to their inquiries, to be challenged and reconstructed with help from materials, peers and teachers.[5] Such reconstruction and organic growth of understanding is holistic and leaves learners more empowered to continue their own learning. They will more easily accommodate to the countless challenges they will meet throughout life than by learning the schizophrenic behaviour taught in most African classrooms.

## Inquiry begins with phenomena

To me a powerful reason to promote inquiry science in Africa is that science starts from inquiry into phenomena, and we are rich in phenomena.[6] The elegance of those such as Newton, Hooke, Faraday and Darwin lies not so much in their discoveries, that have been modified by others standing on their shoulders, as in the ways they persuaded the phenomena of nature to reveal their secrets. The same phenomena, and thus opportunities to acquire similar skills of persuasion, are available to every child in Africa.

Since the start of the 20th century, science has become more opaque.[7] Delving into the particle has made science less accessible to most of us, whatever culture we belong to, both in terms of being able to understand and to practice it. Funding cuts, such as the cancellation of the accelerator that was to have been built in Texas, has American scientists bemoaning the death of physics in the United States. An increasing preponderance of technical black boxes and absence of phenomena in industrial cities may be as important a contributing factor to a threat to science in the United States, and to a growing interest in non-Western ways of knowing. Both old and young are being deprived of the tools whereby they learn and extend an understanding of their trade. In Africa our economies cannot support development of high science, but with our overabundance of phenomena and problems we should be a rich training ground for those with inquiring minds.

## Inquiry alters classroom power relationships

Traditional power relationships shift in classrooms where the touchstone of understanding is the ability to question nature effectively. The teacher's role becomes that of challenging students' understanding as opposed to dispensing knowledge. A key implication of such a change of patterns of authority is that students can challenge each other as well as the teacher in a negotiation of understanding. In such classrooms all children become active participants, including girls.

Many AFCLIST-supported projects have spontaneously commented on an increased interest by girls during inquiry science lessons. I have talked with senior teachers in a secondary school in Malawi who were sceptical when a junior member of staff introduced inquiry science to their school. They reported claiming that traditional behavioural norms would mitigate against questioning, exploring and open discussion. After about a term, the same teachers said they had completely changed their position after noticing that students — particularly girls — participating in inquiry science had become discussion leaders in other subjects.

Basic changes in classroom authority patterns probably achieve more than any other changes, except change in attitude to women, to promote increased participation and performance of girls in science (Erinosho, 1993). Society, not girls, is the problem. However, other changes can contribute to making girls feel more positive towards the sciences. *Action Magazine* pays particular attention to girls as active participants. The leader of the gang in 'Spider's Place', a South African television series, is a girl. Special compensatory programmes, such as the clinics organized in Accra, Ghana, by La Mansaamo Kpee, can have an impact. Examiners can set items within a female culture. But until teachers and parents are aware of the extent to which traditional classrooms render both girls and boys invisible, and until there is authentic science learning worth experiencing, little will change for girls.

I believe that a shared vision and passionate faith in science as inquiry is a necessary factor for change. I equally believe that all parties concerned in the educational process must participate in the development of the vision. We must stop thinking, talking and acting solely in terms of curriculum developer, educational planner, funder, university researcher, examinations and assessment expert, teacher and parent. We must stop expecting radical change and instead accept incremental progress. We must stop thinking in terms of pre- and in-service teacher education and think of continuous professional development. We must begin to see ourselves as equal partners in change, each with something to offer and each with a great deal to learn.

## Teacher education

I discuss teacher education with reluctance. We all need educating and we need an education that continues throughout our professional lives. To focus on teachers alone may be yet again to blame the victim. To talk in terms of pre- and in-service education may be to accept a dichotomy that is self-limiting. We all need an education that provides continued opportunities.

- We need to experience science as inquiry. We need an experience sufficiently extended and focused to motivate us to continue inquiring. We need an education that gives us the confidence and skills to inquire and that puts us in touch with our own creativity. The experience must be broad enough to enable us to be comfortable when inquiry takes us into realms such as physics, chemistry, biology, environmental studies, mathematics and technology. We do not need more so-called background knowledge in any of these realms — a lifetime is inadequate to master a small percentage of the knowledge in any one. We do need to know how such knowledge is generated and how to evaluate it in terms of our own lives.
- We need to be reflective about our own learning. We need to be fully aware of the stubbornness with which we cling to mental constructs in the face of conflicting evidence and of the difficulties we experience in reconstructing our understandings. If we are more analytical about what we know and how we know it, we might be prepared to be less dogmatic, less certain about our certainty, and more prepared to consider the contributions of others.
- We need to work with others, be they children, teachers or colleagues, in ways that encourage their reconstruction of how they understand the world. We need to learn how to work as facilitators and organizers of learning rather than as authoritative teachers of knowledge. We all need continuously to sharpen our skills as negotiators.
- We need to work, with the support of others, at the cutting edge of promoting change. We need continual educating in how to identify strengths and weaknesses within individuals, institutions and systems, in identifying bottlenecks, and in building alliances to overcome them. In short, we need educating in the politics of change.

## Participatory change models identify strengths rather than weaknesses

Participatory change models search for strengths to build on; projects search for weaknesses to fix.[8] Participatory change models are awkward for funders since their objectives are ill defined, they evolve as people evolve, and because there is no time limit on human development. Participatory models are awkward for governments and centralized bureaucracies because as participants gain confidence and become skilled visionaries rather than technocrats, they increasingly question authority.

## SUSTAINING PARTICIPATORY APPROACHES TO PROMOTING INQUIRY

Maintaining school systems in Africa costs countless billions year after year, funds spent on promoting change being virtually nonexistent. Traditional ways have acquired a considerable momentum. Altering direction will require a considerable force applied over years, even if leverage points are cleverly identified. Only by re-directing and harnessing the momentum will change become possible, and doing so implies mass participation. However, the critical mass required for take-off should not be underestimated.

### The role of organizations such as AFCLIST

The role of organizations such as AFCLIST in the development of science and technology education in Africa is captured in the following quotations.

> The relative lack of genuine African input into the formulation of development paradigms separates modern African experience from that elsewhere in the world, and it accentuates the fact that the main challenge for development is to increase the capacity of African entities to analyze past experiences and to formulate new strategies for a better future. (Delgado, 1995)

> In Africa, professionals with a vision of science education have become few and isolated. As the Kenyan experience demonstrated, they are vulnerable to decisions taken on grounds of political expediency, to donor pressure and repeated waves of solutions, to economic hardship and other factors. Yet without vision I see no future for science education on the continent. We can all contribute to building such vision, as recent experiences of projects supported by AFCLIST [have] demonstrated. To do so, I suggest that whatever our position, we must maintain contact with classrooms. We need constant feedback from the enthusiasm, joy and creativity of children when they are permitted to become involved in learning. In the words of the National Advisory Committee in India, recently appointed by the Ministry of Human Resource Development, we can all work to lessen the burden on learning so prevalent in educational systems throughout Africa and re-introduce joy. Whether we are curriculum developers, inspectors of schools, examinations officers, research workers or teachers we can work with others to promote inquiry learning. We can do so within the constraints of Africa since there is no shortage of children to work with, phenomena for inquiry and problems to be solved. Yet the development of such vision needs nurturing. Key sustaining elements include professional fellowship and networking between pockets of effective practice; commitment over longer periods than the lifetime of projects; disinterested professional feedback and support; and funds to facilitate innovative work. Providing such an environment itself requires vision; an African vision so committed to entrenching quality science education in our schools that

> nothing can erase it. We must expect change to be slow and incremental; we must be prepared to start from small pockets of excellent practice networked to slowly gain momentum. No donor could or should bear responsibility for developing such vision, though they can support us in its implementation. We must do so ourselves. Only organizations such as AFCLIST can sustain the vision and provide a safe haven in a continent that remains unpredictable.
>
> However, Africa has seen organizations such as AFCLIST blossom, flower and die. The Science Education Programme for Africa, the African Curriculum Organization and the Federation of African Science Educators are three such organizations. In one case, though the organization is still officially registered, 'financial mismanagement' was a major cause of its demise. Its African constituency permitted the death as it permitted those of the other organizations named. When we are criticized, we cry lack of funds or intellectual imperialism and rarely look within. Perhaps we should do so. Some teachers continue to work in dedicated and inspired ways. They have no choice but to engage with their own constraints and realities. Their biggest resource is the constant inspiration they receive from the creativity of children who remain the largest untapped asset in Africa. [We] researchers and experts should seek our sustenance from the same source. 'It's just a matter of struggling,' as Samuel Githinji said 20 years ago. To evolve effective ways of working with teachers and children to unleash their potential would put us right back on the cutting edge of international science education research and practice.

I had thought that such work could no longer be found. However, recently, the daughter of Dr Eddah Gachukia recounted how she had spent days searching for a lively primary school science class to video-tape. She was driving home in despair, passed a primary school and thought she might as well try one last time. She encountered a school of Githinji's quality and her excitement when she described it was infectious. The teacher responsible had attended a five-and-a-half-day in-service course organized by the Kenya Institute of Education during the mid-1970s. In the course of discussion with the teacher, head and deputy head, Eddah's daughter asked about the involvement and performance of girls. She reported that they answered with some puzzlement that girls were thought to have problems, showed her charts detailing their excellent performance, and reported that a girl from their school had had the best result of any girl in the district in the primary school leaving examination in 1994. I find it awesome that some teachers can continue to inspire children under current constraints.

## NOTES

1 A slide-tape presentation of the same name made by the Kenya Institute of Education gives a lively picture of Githinji's classroom.

2 By 1971, 40 % of America's high school teachers had participated in summer institutes funded by the National Science Foundation, where they were exposed to courses developed by the curriculum teams of the 1960s. In 1978 Stake and Easley report that the inquiry approach, hands-on student experimentation and student-initiated discussion are not in common use in most schools (Stake & Easley, 1978).

3 For a discussion of science as inquiry and its implications for teaching, see *Standards for Science Teaching and Professional Development of Teachers of Science.*

4 Concern with elitism and 'toughness' seems to emerge from a search for mathematical relationships; with the formulae. Such concern may be misplaced. I have seen Form I pupils in Zanzibar in their own puzzlement, not at the instruction of the teacher, struggle towards quantifying the relationship between mass and volume to better understand the phenomenon of sinking and floating into which they were inquiring. To ban quantification from science on grounds of elitism seems to me to be underestimating children.

5 For a discussion of science education and constructivism see, for example Duckworth, E. *The Having of Wonderful Ideas and Other Essays* (New York Teachers College Press, 1987); Driver, R. Constructivist Approaches to Science Teaching (Paper presented at the University of Georgia Mathematics Education Department).

6 David Hawkins provides powerful arguments that science begins with inquiry into phenomena in *Messing About in Boats, An Elementary Science Reader* (Newton, Ma: EDC, 1969).

7 It is interesting to note that this turning point in the sciences coincided with a demise of public education efforts. Fewer of the great 19th century exhibitions vaunting the products of science and technology were held. Museums such as Urania in Berlin, which invented hands-on displays, were closed. Not until the mid-1970s was there similar concern with mass education through interactive museum displays of basic phenomena.

8 In *Science Education for the 1990s,* Mark St John gives a list of 27 ways in which project-based change models differ from systemic change models. Participants at the Wingspread Conference, each of whom had decades of experience in curriculum change, showed a clear preference for the systemic change model.

## REFERENCES

Delgado, CL. 1995. *Africa's Changing Agricultural Strategies: Past and Present Paradigms as a Guide to the Future.* International Food Policy Research Institute

Duckworth, E. 1987 *The Having of Wonderful Ideas and Other Essays on Teaching and Learning.* New York: Teachers College Press, p 64

Erinosho, Sheila Y. Preferences of Nigerian High School Teachers for Modes of Assessment. *Studies in Educational Evaluation;* 19(4) pp 439–45

Hawkins, D. 1994. A personal response to *Standards for Science Teaching and Professional Development of Teachers of Science*, unpublished

Jegede, OJ. Collateral learning and the eco-cultural paradigm in science and mathematics education in Africa. *Studies in Science Education,* 25, pp 97–137

Keogh, M & Salamon, C. 1994. Insights from genetic theory towards a theory of educational change. A paper presented at the Southern African Association of Researchers in Maths and Science Education, 27–30. January 1994. Durban, South Africa

Lapp, DM. 1980. *The Improvement of Science and Mathematics Education in Less Developed Countries.* Institute for Scientific Planning and Technological Cooperation

Lapp, DM. *The State of School Science: A Review of the Teaching of Mathematics, Science and Social Studies in American Schools, and Recommendations for Improvements.* National Research Council

Lapp, DM. 1983. *Basic Science Education in Sub-Saharan Africa.* United States Agency for International Development

Moses, RP *et al.* 1989. The algebra project: Organizing in the spirit of Ella. *Harvard Educational Review,* 59(4). November, pp 27–47

Shankerdass, S. 1993. *Developing Strategies.* A multimedia pack developed for AFCLIST

St John, M. 1991. *Science Education for the 1990s: Strategies for Change.* Inverness Research Associates. Sponsored by The Johnson Foundations, Inc, Racine, Wisconsin

UNESCO (United Nations Educational, Scientific and Cultural Organization). *Teaching and Learning in Science & Technology*, vol 2. Paris: UNESCO

World Bank. 1994. Report on Education for All. Washington

# 4

## Who shapes the discourse on science and technology education?

*John Volmink, Centre for Advancement of Science and Mathematics Education (CASME), University of Natal, Durban, South Africa*

### ABSTRACT

This chapter identifies dominant trends or discourses in various aspects of science and technology education in African countries. These are shaped and determined by particular interest groups with conscious or unconscious agendas. The chapter examines who shapes the discourse of science and technology in Africa and analyses who and how groups, including science and technology educators, scientists and technologists, industrialists, education policy makers, economists, politicians, researchers, donors, the World Bank and foreign aid, shape discourse, practice and policy in science education.

### INTRODUCTION

Discourses are created by people: they are social artifacts and subject to change. Discourses are rules that govern how we create meaning and ascribe value. Within a given field, the discourse is shaped by the powerful: those who have a voice. As a community of African science educators, we must face the question: Who shapes the discourse on science and technology education? In seeking an answer, we should not try to find 'culprits' but try to understand processes and structures.

In this chapter, I deal with the question of control in science and technology education. The issue of hegemony is political, hence my focus on ideology rather than pedagogy. Much of the discourse on science and technology education is shaped in the classroom by teachers and learners, but since this discussion is taken up by other contributors (cf Savage: Chapter 3), it does not form a major part of this chapter. In addition to ideology, I also consider the role of epistemology, namely what counts as knowledge in science and technology education and who decides what knowledge is worth knowing. I believe it is crucial that we understand how structures that legitimize oppressive forms of control are produced and reproduced, so that collectively we can take appropriate action to counter them. In an attempt to map a terrain for debate, this chapter focuses on various dimensions of the discourse on science and technology education. To help develop counterhegemonic strategies, I explore the areas of ideology, epistemology and structures to find an explanation of how they shape the discourse.

I do not look for the 'who'. I find it hard to accept a 'conspiracy theory'. We all shape the discourse on science and technology education in our spheres of influence. There are some who, by deliberate and sometimes devious means, have acquired power and are unwilling to give it up: they refuse to question their assumptions of entitlement. There are others who, because of uncritical practices and unquestioned assumptions, wittingly or unwittingly participate in these modes of organization. We should interrogate these power relations to become aware of their pernicious effect in our own contexts.

As a community of African scientists and science educators, we see that many of the key issues in the international debate on science and technology education do not include our lived experience but have been defined in another place at another time. While recognizing the value of learning from other contexts and experiences, we should no longer give our uncritical allegiance to every wind that blows from the North. This does not mean that we wish to delink ourselves from the rest of the world, but we do seek recognition of our thoughts and perspectives. Such recognition would contribute significantly to science and technology education, discourse about which should be reinterpreted within the context of the global village.

An underlying assumption of this chapter is that science education is an educational rather than a scientific or technological endeavour. Most science educators recognize that science education lies at the confluence of other fields, such as education, anthropology, psychology, politics. One could argue for the primacy of any one of these fields as they pertain to science and technology education and I will state my biases. I construct my argument recognizing the influence of scientists on the early development of science education. However, science and technology education has emerged as a separate discipline and is no longer seen as merely a subdiscipline of science and technology. In order to gain a perspective of science and technology education in our own contexts, we need to test our assumptions and re-examine the historical development of our current practices.

## SCIENCE, TECHNOLOGY AND THE IDEOLOGY OF DOMINANCE.

Strong hegemonic forces impose a certain view of science on us all. Our schooling has encouraged us to accept that the traditional science curriculum embodies powerful knowledge and eternal truths that should be learned in a catechistic fashion. Furthermore, many see this knowledge as infallible and universal. Scientists portrayed their discipline as a set of bounded meanings, with a well-defined, culturally neutral and value-free inside, and a large outside. Fortunately many scientists, and certainly the best, do not hold to this view. However, some have generated an image of science and of themselves that has provided them with a virtually unchallenged right of passage. As Feyerabend (1981: 157) says: 'In society at large the judgments [sic] of scientists is received with the same reverence as the judgments of bishops and cardinals was [sic] accepted not too long ago.' They, as high priests, now have the responsibility and sole right to decide what should be included and what should be excluded. An impenetrable wall has therefore been erected around what has become known as science and the scientific method and it is this wall that inhibits innovation or any radical departure from the canonical school curriculum. It is clear, however, that much pressure is being exerted on this wall by, among others, non-traditional scientists, science educators, philosophers of science, anthropologists, educational ethnographers, sociologists and just plain old folk doing their everyday work. Recent work under the name of 'world-view' has added much to the growing evidence illustrating the non-universality of formal science.

Feyerabend (1981) argues that he wants to defend society from all ideologies, science included. He asserts that although science was once in the forefront of the fight against authoritarianism and superstition, it has become rigid and has ceased to be an instrument of change and liberation. It has become an ideology and there is nothing essentially liberating in science or in any other ideology. He says (1975): 'Modern science overpowered its opponents, it did not convince them. Science took over by force, not by argument. Science would have been impossible without dogmatism.' Yes, scientists convince, persuade and overpower each other with evidence and argument that uses evidence. It is the assumption that the scientific method is the most effective and powerful approach that comes under special attack from Feyerabend. I share much of his concern about the role of science as an ideology because of how it dominates our world-views and how, as an ideology, science provides frameworks for action.

Science tends to classify, label, assess and measure all that is human and non-human. Science becomes driven by a desire to control and to dominate, and thus to exercise power over others and over nature. Science and technology routinely divorce fact from value, and favour fact. This can lead to devaluation and marginalization of people and to the creation of an otherness. Our uncritical use of science and technology has polluted our lakes, poisoned our rivers, made holes in the ozone and has acid rain falling from the clouds. We have developed a capacity to destroy ourselves many times over and everywhere see crime and unrest and disease and war. In what way is this a better world? Science has been described as 'the most distinctive enterprise of Western civilization in the 20th century'. Yet we should not view science and technology as panaceas to our problems. We must recognize their limitations as well as their benefits in relation to technical, social and economic development.

An absence of scepticism of the dominant view of science has led to uncritical adoption of Western technologies and methods. In his book *Machines as the Measure of Men*, Michael Adas (1989) explores the role of science and technology in shaping ideologies of Western dominance. Early European expansionists and anthropologists were more preoccupied with material expressions of culture than with modes of organization. Their interest in the material and technological accomplishments of Africa, India and China were not mere academic exercises — they were expressions of power relationships. Descriptions by European explorers of African tools, Indian modes of transport, Chinese timepieces and so on, served to shape European perceptions of their own scientific and technological superiority. These perceptions provided a better justification for their 'civilizing mission' than religion, since the superior science and technology would bring economic and cultural advancement. Adas describes it thus:

> . . . evidence of scientific and technological superiority has often been put to questionable use by Europeans and North Americans interested in non-Western peoples and cultures. It has prompted disdain for African and Asian accomplishments, buttressed critiques of non-Western value systems and modes of organization, and legitimized efforts to demonstrate the innate superiority of the white 'race' to the black, red, brown, and yellow.
>
> Throughout the centuries . . ., European judgments about the level of development attained by non-Western peoples were grounded in the presuppositions that there are transcendent truths and an underlying physical reality that exist independent of humans, and that both are equally valid for all peoples. Further, most of the travelers, social theorists, and colonial officials who wrote about non-Western societies assumed that Europeans better understood these truths or had probed more deeply into the patterns of the natural world that manifested the underlying reality. (Adas, 1989: 6)

Observations only served to show that 'European modes of thought and social organization corresponded much more closely to the underlying realities of the universe than did those of any other people or society, past or present'. (Adas, 1989)

Adas argues that, while Europeans became increasingly dissatisfied with using scientific and technological development as gauges of human worth, particularly after World War I, Americans have continued to do so.

However, I believe there is enough evidence that the ideology of dominance, although it has become more subtle, remains pervasive and universal. As an example, Reiss (1993) cites the recent British Broadcasting Corporation (BBC) World Service broadcast, 'They made our world'. The expectation from a World Service broadcast is that it would have an international flavour. Instead, the series, and the accompanying book, focuses on the work of Bacon, Newton, Priestly, Lavoisier, Faraday, Maxwell, Lyell, Darwin, Mendel, Jenner, Pasteur, Fleming, Watt, Stephenson, Bell, Edison, Write, Ford, Roentgen, Marconi, Baird, Baekeland, Turing, Einstein, Rutherford, Oppenheimer and the Manhattan Project. None are from outside Western Europe and the USA, and all are male. Reiss points out that since there are no absolute or universal criteria by which scientific excellence can be infallibly judged, who is considered a great scientist depends on one's point of view. As Faseh (1993) so rightly says, 'Hegemony is not only characterized by what it includes but also what it excludes: by what it renders marginal, deems inferior and makes invisible.'

Our blinkered vision has stultified and distorted our thinking and the world is poorer for it. We need to realize and understand the value of all ideas and perspectives and not just of a select few. We would do well to heed the advice of scientists such as F David Peat (1994), a theoretical physicist, who, after a long encounter with the Native Americans, says, 'Perhaps the time has at last come when we can simply sit down, listen, and come-to-knowing. Maybe, as the millennium reaches its close, we can all engage in a ceremony of renewal that will cleanse earth and sky. Maybe the time is right.'

As Adas puts it:

> Less arrogance and greater sensitivity to African and Asian thought systems, techniques of production, and patterns of social organization may have enhanced possibilities of evolving alternative approaches to development — approaches that might have suited Third World societies better than the scientific–industrial model in either its Western or its Soviet guise. At the very least, the first generations of Western-educated leaders in the newly independent states of Africa and Asia would have been more aware of the possibilities offered by their own cultures and less committed to the industrialization that most viewed as essential for social and economic reconstruction (Adas, 1989: 16).

I am not simply making a plea that the views of other cultures be integrated into the global hegemony. Integration would imply a fundamental challenge that we critically question whether the Western scientific–technological culture is a model that should be imitated by the rest of the world. Neither, however, should the model be discarded simply because it is Western — this would be a form of counter-hegemony that I find unhelpful and reject. Being critical means that we are aware. This awareness assists us to recognize the delicate web of connections between us all, and to identify those forces and power relations that prevent us from deriving benefit from the web. If our vision is what Longino (1993) refers to as a 'transformative critical discourse', where there is equality of intellectual authority in the science and technology education community, then we have to accept her suggestion that 'no section of the community, whether powerful or powerless, can claim epistemic privilege'.

## DOMINANCE AND THE SCIENCE CURRICULUM

Several chapters deal with the historical influence of the overseas curriculum reform movement on African science and technology curriculum change. Any discussion of curriculum and curriculum change raises important questions. Pertinent to this discussion is the question of control: Who is involved in the discourse on science and technology education, who owns it, and in whose interest is the discourse perpetuated?

We should distinguish science as a discipline from science as subject matter. In the process of determining subject matter, we need to consider the range of interest groups with contending views as to what knowledge is worth teaching.

Williams (1961) distinguishes three ideological groups in industrialized societies that influenced education in the past and continue to do so. He calls the first group industrial trainers. This group represents the merchant, managerial and some professional classes, who share the aim of education as preparation for work. Their interests are narrow and utilitarian. Their social concerns do not go beyond instilling basic skills and obedience. As science educators, they stress drill-and-practice and other forms of rote learning and assessment.

Williams refers to the second group as old humanists. They represent the elite who value the cultured, well-educated person. Old humanists place great value on the transmission of the cultural heritage, and see the aim of science education as producing the new generation of pure scientists.

Finally, Williams identifies public educators as a group of radical reformers concerned with democracy and social equity. Their aim is 'education for all' to empower the working classes so they can participate more fully in the prosperity of modern industrial society and in its democratic institutions. They want to see science students being encouraged to critically examine the use of science and technology in society.

Paul Ernest (1991) introduces two additional ideological groups, namely the technological pragmatists and the progressive educators. Technological pragmatists represent the interests of industry, commerce and public sector employers. They value practical skills and technological progress. In addition to bringing technology challenge to science education, they go beyond industrial trainers in that they encourage a broad range of skills such as communication, problem solving and so on. Technological pragmatists emphasize the utilitarian aspects of science and technology without necessarily questioning their nature.

Progressive educators, on the other hand, are romantic, liberal reformers, whose emphasis is more child centred than that of public educators. They are the modern representatives of a tradition whose proponents have included Rousseau, Montessori, Dewey and Piaget. In science classrooms they emphasize creativity and self-expression.

Of course, these groups do not constitute an exhaustive list of stakeholders, but illustrate that there are always contending ideologies for dominance in the curriculum discourse. Ernest (1991) documents how these groups, vastly unequal in terms of their power, impacted on the British national mathematics curriculum. It is clear the impact made by each group was commensurate with its relative power, that varied from the overwhelmingly powerful industrial trainer through the powerful old humanist–technological pragmatist alliance to the marginalized public educator.

In addition to these stakeholders, there are other influences on the curriculum discourse. One such influence is 'international trends'. Sylvia Ware (1992) summarizes these influences as coming in two waves of reform.

> Thirty years ago, the United States (with the NSF-supported 'alphabet' curricula) and the United Kingdom (with Nuffield science) began the reform of primary and secondary science curricula that was to spread, often with few modifications, to the rest of the developed and developing world. For this first wave of reform there were two generally accepted purposes: the initial training of the next generation of scientists, and a belief that science knowledge was in some way important to the intellectual development of all students. The second purpose was soon to become subordinate to the first. (Ware, 1992: 8)

Ware (1992) summarized a 'second wave of reform' under the rubric of 'science (and/or technology) for all'. All students are targeted for science instruction, even though most of them are unlikely to become scientists. Essential content is redefined so that science is taught and learned from and within its cultural context. Second wave courses are less elitist, learners are more active, teachers are more open, and content focuses more on societal issues than on disciplines.

**Table 4.1: A comparison between the first wave and the second wave of science curricula**

| First wave | Second wave |
|---|---|
| Preparation for science career | Science for all students |
| Generation of science knowledge | Application of knowledge |
| Focus on the discipline | Focus on societal issues |
| Broad coverage of content | Less content = more learning |
| Science on the lab bench | Science in the community |
| Building of conceptual models | Personal decisionmaking |
| Mastery of content | 'Ownership' of content |
| The teacher as a lecturer | The teacher as a manager |
| Classwork as a unit | Students work in groups |

This second wave reform took place in the industrialized countries and has been taken up only in limited ways elsewhere. Ware (1992) puts it as follows:

> Much of the momentum of the second wave of science curriculum reform can be credited to the UK Association for Science Education, which in 1981 published 12 readers in the series *Science in Society*. Other countries were also involved in early reform, including the United States, Canada, the Netherlands, Thailand, Australia, and New Zealand. While second wave reforms are now being implemented in many countries, this movement has,

so far, had a fairly limited impact on science curricula in much of the developing world. Particularly at the upper secondary level, the first wave courses still predominate, minus the 'discovery' approach to laboratory work. At the lower secondary level, the spread of integrated science can be considered a bridge between the first and the second waves of science curriculum reform. (Ware, 1992)

Table 4.2 illustrates Ware's observation that, though currently limited, some countries in Africa are adopting an integrated approach to science and technology teaching.

**Table 4.2: Curriculum emphasis for selected countries**

| Country | Grade | Curriculum emphasis |
|---|---|---|
| Bahrein | jr sec | Single sciences: academic (1987) |
| Bangladesh | 9–10 | Academic |
| Barbados | 10–11 | Integrated and single science schemes: themes, concepts |
| Bhutan | 7–10 | Academic (1989) |
| Botswana | 8–9 | Integrated science: academic with some societal topics (1987) |
| Ghana | 10–12 | Academic science (science concepts) (1990) |
| India | 9–10 | Thematic with concepts, work relevance; also course on work experience (1988) |
| Jordan | S1/S2 | Single science: academic |
| Korea | 7–9 | Science: academic |
| | 10–12 | Single science: academic |
| Malta | 9 | Integrated: concepts |
| New Zealand | 7–11 | Science: relevance, skills, concepts |
| | 12–13 | Single sciences: academic (1990) |
| Nigeria | 7–9 | Intergrated science: academic with themes, social topics (1985) |
| | 10–12 | Single sciences: academic |
| Pakistan | 9–10 | Academic |
| Philippines | 10–11 | Tertiary (includes science skills) |
| Sierra Leone | 7–9 | Integrated science: concepts with themes |
| Trinidad | 10–11 | Science: values, environment |
| Zimbabwe | 7–9 | Academic (1989) |
| | 10–11 | Academic (1990) |

In post-apartheid South Africa, and to a lesser extent in other African countries, the discussion increasingly centres on providing 'science for all' rather than for the privileged. Under the banner of 'preservation of standards', the old humanists — the elite who value the cultured, well-educated person — line up against the public educators whose concerns are the promotion of equity. I can envisage a scenario where we may have to give up ownership of the disciplines we guard so jealously, and adopt a collaborative, coherent and integrated approach to education. However, though I support the desirability of such an approach, I realize we must address not only ideological implications, but also the conceptual and practical difficulties involved in introducing such sweeping changes. Many can be described in terms of paradigm shifts, and while I accept that we should recognize the opportunity for change created by paradigm shifts, we should also recognize that they cannot be imposed. The status quo by definition works and has its own momentum. Even those excluded from the discourse desire access to an education that served the elite so well. Thus, any paradigm shift must come from a widely shared perspective. Indeed, it may be counterproductive to suggest radical change without taking into account the extent that the vision is shared and, equally important, the systemic implications of change. We should perhaps work towards a gradual transformation of the curriculum and instruction to support the goals of our societies, rather than polarizing power structures from an inadequate base. However, as science educators we must always realize that we are more than 'technicians'; that we are part of the discourse and power structure; and that our actions and work have long-term social implications.

## STRUCTURES THAT PERPETUATE THE STATUS QUO

The form of the discourse is maintained through institutional and structural arrangements, some more subtle than others. Consider, for example, the resourcing of science and technology education practice and research. Governments and donors have agendas that entrench certain practices. Funding policies do not only determine the configuration of teaching spaces and teacher–pupil ratios, they also fundamentally affect the form of the discourse within classrooms. Often policies are not informed by research, but are assertions made by politicians, bureaucrats or donors. It is therefore notable that little money is available for policy-related research in science education that could assist in placing these decisions on an informed basis.

Funding of science education research tends to favour those whose agendas support the status quo. Consequently, issues such as policy development and analysis are underrepresented in the literature. I was therefore particularly pleased to see that the *International Journal of Science Education* recently devoted a special issue (volume 17, number 4) to policy and science education. On the other hand, programmes such as the International Education Association (IEA) studies on achievement and the Third International Measurement of Mathematics and Science (TIMMS), that serve only to provide an international comparative construct, remain heavily funded.

 

In many African countries, it is difficult to find funds for effective, system-wide innovation in science and technology education. The salary dilemma of university-based academics encourages them to accept international consultancies that have their own agendas rather than engage in nationally sponsored research. However, there are encouraging counterexamples, such as projects supported by the African Forum for Children's Literacy in Science and Technology (AFCLIST), funded by the Rockefeller Foundation.

The most powerful determinants of discourse are those structures and institutions that provide a voice. Mass media, publishing houses and other vehicles of communication play a crucial role in defining issues and creating champions. Science and technology in Africa and other parts of the non-Western world are poorly represented in this communications network. For example, a disturbing article by W Wayt Gibes entitled 'Lost science in the third world' appeared in the August 1995 edition of *Scientific American*. In this article the author details ways that 'Third World' scientists are marginalized by international journals. Although these countries account for 24,1 % of the world's scientists, most leading journals publish far smaller proportions of articles from these regions. In 1994, the journal *Science* accepted only 1,4 % of the articles submitted by authors from developing countries. By contrast, the same journal published 21 % of all articles submitted from the United States.

Over the three-year period since 1991, the number of articles submitted from developing countries to *Science* has doubled, yet the acceptance rate has remained the same. Floyd E Bloom, the editor of *Science*, is quoted as saying about Third World scientists: 'If you see people making multiple mistakes in spelling, syntax and semantics, you have to wonder whether when they did their science they weren't also making similar errors of inattention.' It is interesting to note, however, that non-English-speaking European scientists, such as German and French, have a much higher acceptance rate than Indians. This provides clear evidence of bias and insensitivity.

Then there are epistemic structures such as constructivism that play a significant role in shaping the discourse in science and technology education. Constructivism has provided a powerful challenge to the dominant positivistic ideology. Whereas behaviourism is grounded in a mechanistic and reductionist world-view, constructivism finds its origin in a world-view that is holistic and dialectical. Kauffman (1975) argues that a constructivist approach to knowledge forces us to re-examine our theories of human nature since they correspond structurally to different political ideologies. I was originally attracted to constructivism because it provided me with a different perspective from which to value my own work as well as that of my students. It provided an epistemological base to take into account different points of view within a relativistic framework.

But constructivism has become the new orthodoxy in mathematics and science education. It has assumed a dominance in all fields of education. Any current discourse on learning and teaching in science and technology education is affected by the constructivist perspective that must feature at science education conferences all over the world.

 

A growing critique of constructivism is emerging. In a paper entitled 'Constructivism as a liberal bourgeois discourse', Robyn Zevenbergen (1995) gives a brilliant analysis of the role that constructivism plays in the discourse on mathematics education, drawing on the work of Pierre Bourdieu. Transposing her argument to science education, it would go as follows: Science education, like any other field, operates within its own set of rules and logic. These define and constrain what is valued, and produce intellectual legitimacy for participants in the field. Within a field, participants establish credence by amassing 'capital', while complying with the unspoken rules and logic. This conveys power and status. Those who amass more capital are able to speak with greater legitimacy than others. Within science education, constructivism can be seen to be a form of symbolic capital and those who amass it can convert it into other forms of capital such as economic capital (higher salaries, research grants and so on) or institutionalized capital, such as prestigious appointments. Those lacking the desired capital within a field are not given the right to speak and are relegated to marginal status. Discourses that are critical of these practices are not given legitimacy and this leads to intellectual censorship.

Such intellectual censorship applied to other issues in science education, such as the 'misconceptions industry' of the previous decade. Therefore my criticism is not of constructivism or of misconceptions research, but of the rules and logic that shape the discourse. Zevenbergen and others, however, do raise criticisms about the shortcomings of constructivism. Pertinent to this discussion is an awareness that constructivism and other leading theories play a fundamental role in shaping the discourse on science and technology education. The degree of consideration given work by journals, conferences and funders becomes linked to the extent that it shows evidence of familiarity with the current, dominant idea.

Furthermore, our belief structures often buttress incongruities in science and technology discourse. Faseh (1993) argues that, within hegemonic contexts, 'others' are often accorded honorary status. Because of the sense of self-worth and status derived from this vicarious participation, Western-educated intellectuals from Third World societies, he argues, tend to overvalue symbolic power such as titles, degrees and access to prestigious institutions, journals and awards.

Drawing on his own experience as a Harvard-educated mathematician from Palestine, Faseh illustrates how this realization affected the way he saw his own contribution to his community. While his own validation group was the 'international community', he undervalued the mathematics of his mother's sewing. His work had wide recognition while his mother's only had local impact. He articulated his ideas in ways the international community could understand; she could not. The question raised is not which is better, but rather how can these different ways of knowing be brought together? It cannot happen as long as we undervalue other legitimate ways of looking at the world, or when interest in situated knowledge, ethnoscience and ethnomathematics remains limited within the discourse of science and technology education. Even those who are oppressed by this ideology participate in their oppression and exploitation by leaving these practices unchallenged.

The power imbalance in science and technology education does not lie only along a Third World/First World divide. Institutional arrangements within a country can be as oppressive, and are often based on race, gender or class. In South Africa, for example, generally white, male, middle-class scientists and educators based at prestigious universities shape the discourse on science and technology education (Reddy, 1995). This group makes the rules for research and pedagogic practices and policies, supervises science education graduate students, and has the power to force them to comply. This group, more than any other in the country, perpetuates the status quo in science and technology education. Such class struggles generally control social structures and result in power being given to very few people. This is as true in African countries as elsewhere in the world.

Finally, officials and bureaucrats are an important group in the discourse of who controls science and technology education. Yet many become jaded and cautious, follow their routines, rarely question the appropriateness of what they do, and have lost any inclinations they may once have had to innovate. Through inaction, they represent a strong force in maintaining any status quo.

## COUNTERHEGEMONY: IS THERE ANY HOPE?

Alexander Colander's story related in *Teaching Elementary Math and Science* is well known. He was asked to referee a dispute between a professor and a student about the grading of an exam question. The professor was about to give a zero for the answer while the student claimed he deserved full marks.

Both agreed to an independent arbitrator. The exam question was: 'Show how it is possible to determine the height of a tall building with the aid of a barometer.' The student's answer went as follows:

> 'Take the barometer to the top of the building, attach a long rope to it, lower the barometer to the street, bring it up measuring the length of the rope. The length of the rope is the height of the building.'

Colander told the professor that the student had a strong case, but agreed that full marks in a physics course should indicate competence in the subject. Both parties agreed to a retest. The student was given six minutes to answer the same question, and was told that his answer must demonstrate a knowledge of physics. Colander described what happened.

> At the end of five minutes he had written nothing, I asked him if he wanted to give up. He said, 'No. I have many answers to this problem, I'm just trying to decide on the best one.' I apologized for interrupting and he began writing feverishly. His answer was, 'Take the barometer to the top of the building, lean over the edge, drop it and using a stop watch time how long it takes to hit the ground. Then using the formula s = 3Dgt2, you can calculate the height of the building.' At this point I asked my colleague if he wanted to give up, and he awarded the student almost full marks. On the

> way out of my office I remembered the student said there were many ways to calculate the height of building using a barometer, so I asked him to mention a few. He said,
>
> 'You can take the barometer out on a sunny day, measure the length of the barometer and the length of the shadow of the barometer. Then measure the length of the shadow of the building and by the use of simple proportion you could determine the length of the building.'
>
> 'Fine,' I said, 'any others?' He said,
>
> 'Oh yes, if you want a more sophisticated method, you can tie the barometer to the end of a string, swing it like a pendulum, and determine the value of g at the street level and the value of g on top of the building and from the difference of the values of g the height of the building can in principle be calculated.'
>
> 'Or,' he said, 'there is a very basic method you would like. In this method you take the barometer and you begin to walk up the stairs of the building. As you go up you mark off the length of the barometer on the wall with a pencil. You go all the way to the top, and then you go all the way back down and count the marks and you will have the length of the building in barometer units.'
>
> 'But,' he said, 'the best method is to take the barometer to the basement and knock on the superintendent's door. When he answers you speak to him as follows, 'Mr. Superintendent, here I have a fine barometer, I will give it to you if you tell me how tall this building is!'

Colander's account provides a context within which we can appreciate the complexity of the structures and processes that shape the discourse in science and technology education. The delightful story illustrates the constant struggle at all levels of society between hegemony and counterhegemony.

## Counterhegemony in classrooms

### *How we teach: empowerment of learners*

The professor in Colander's story wanted evidence from the student that he possessed an assumed, well-defined, intellectual capital. This capital was to be accrued in a predetermined way and had to be manifested in a prescribed form. This orthodoxy only calls for reproduction, its intent is to preserve the status quo, and it reduces the possibility of more desirable alternatives. The student refused to participate in this form of domination. He did not grant the professor any 'epistemic privilege'. It would have been easier for him to have done so, but it would have been at the expense of the discourse and of the learning experience of all concerned.

We must question and challenge our own practices. Authority-based teaching and learning must make way for investigative, learner-centred approaches. Teachers must become more open, receptive and reflective, learners more creative and critical. In classrooms at all levels, power must become more equitably distributed. I believe such approaches begin when we act purposefully and with awareness towards understanding and acting on the physical world. Through science we can: (1) structure our experience of the world; (2) understand and transform the socio-political realities that impact on our lives since authentic learning makes us aware of social inequalities and underlying assumptions of social organizations, and (3) create new ideas, perspectives, insights, and models. We cannot accomplish these through approaches such as behaviourism that see students as recipients of information rather than as active participants. As educators, we must all resist the impulse to rush to closure, because this invariably means an end of the dialectical process. Discourse demands that we suspend our need for closure and our craving for a lack of ambiguity. To see this culture operative at a macro-level, we must begin by cultivating it in the classroom. This is exactly the struggle that needs to happen at the macro-level in science and technology education.

### *What we teach: empowerment of teachers*

What we teach in classrooms is as important as how we teach in determining the discourse in science and technology. Society cannot continue to afford the luxury of sending children to school to pursue knowledge simply for its own sake. The socio-economic and political realities on our continent are such that students must pursue knowledge for life. So, school science must become 'science for life', instead of decontextualized, esoteric, abstract and useless knowledge. Science and technology education must become relevant and meaningful.

Historically, science has been defined in Africa by universities. Adherence to a 19th-century model of science inherited unquestioned from the colonial era has for the most part failed to address the development problems facing the continent. Yet science at tertiary level is itself being redefined, as is discussed in chapter 1 above. Increasingly the science community is engaging in a multidisciplinary style of research that is necessary to transform rural and informal sector economies. However, few scientists in Africa engage in demystifying science as do the Sagans and Goulds in the United States. The public image remains that of lab-coated, absent-minded professors working in their physics, chemistry or biology laboratories.

We cannot become transformative by remaining confined to such arbitrary subject-matter boundaries. But can 'integrated science' exist on its own while 'meaningless mathematics' continues to be taught? Increasingly I believe that we should look at a broader programme of meaningful education in Africa, rather than at meaningful science, history, and so on. Curriculum change cannot happen in isolation of the broader educational debate and context.

We should change through discourse rather than by decree or imitation. We need to heed the dangers of centralized control of the curriculum. It is not only a threat to authentic discourse but is a statement that teachers are untrustworthy and need strong governance to keep them in check and to ensure quality. Centralized control does not ensure quality; on the contrary, it destroys it. Science teaching is assessment driven because of centralized control. I have worked sufficiently with teachers all over South Africa and elsewhere to be convinced that they take the quality issue seriously and can be trusted to act in the best interest of students and of society. They feel burdened by the centralized curriculum. The politics of educational change in democratic societies requires involvement, not imposition. Ownership of the curriculum discourse needs to be placed where it belongs — with the people in local communities.

## Counterhegemony in support structures: empowerment of innovators

I am not suggesting that there are no initiatives in Africa that embody criteria of excellence. There are many, and other chapters describe them in depth. The concern of this chapter is to what extent these initiatives, projects and programmes have influenced the discourse on our continent and worldwide. They may be excellent examples of good practice, but while they stand disconnected from each other, their potential to generate an alternative discourse remains limited. One initiative, I believe, stands alone as a beacon of hope. The African Forum for Children's Literacy in Science and Technology — a consortium of African educators, scientists and media specialists — has not only funded approximately 60 projects in over 20 African countries since its establishment by the Rockefeller Foundation in 1988, but has uniquely promoted and facilitated a culture of discourse between its stakeholders. As the project coordinator, Agnes Katama (1995), has put it: 'All manner of partnerships have been critical in the formation of systems of education on the continent, but few can overshadow the web that has slowly emerged as a result of those links between seasoned educators, policy-makers, media specialists and teachers who within the mandate of the Forum have been advocates for the scientific curiosity of the child.' We are deeply indebted to the visionary leadership that has helped this Forum to develop and stand as a sparkling exemplar of counterhegemony on our continent. We need to consolidate and expand it to every corner of the continent. AFCLIST's potential for changing the dominant discourse in science and technology education is enormous.

## Counterhegemony internationally: empowerment of researchers

We need to go beyond counterhegemony. We must also penetrate the First World culture of dominance. If we leave it untouched, it will simply return. We need to communicate the value of our models. One way to do so is through the written word and we have not done this well. We have not held up our ideas to public scrutiny. I am encouraged by the willingness of the editors of journals such as the *Journal for Research in Science Teaching* and the *International Journal of Science Education* to consider contributions from African science educators. But perhaps the time has come for an African journal of science and technology education as a means of communicating with each other and the rest of the world. We should encourage funders such as the United Nations Educational Scientific and Cultural Organization (UNESCO) and the International Development Research Council (IDRC) as well as African governments to support such a venture.

 

The research agenda in science and technology education in Africa needs to be reconceptualized and driven in a different manner. In South Africa, for example, the Southern African Association for Research in Mathematics and Science Education (SAARMSE) was established in 1992 to redress the historical imbalances created by the apartheid legacy. The association has grown from some 40 members to well over 300 within three years. It provides opportunities for building research capacity and for networking across the southern African region. Issues such as gender equity receive special attention with assistance from the Foundation for Research Development (FRD). Similar efforts are being made across the continent, but progress is slow. It is crucial that African science educators see themselves as knowledge producers as well as knowledge users, since research shapes discourse as well as informing practice.

## CONCLUSION

The extent to which 'others' have dominated the discourse in Africa is not, however, the key question. A more important question is: 'If we believe that the discourse on science and technology education in our context must rest with us as the stakeholders, what we are going to do about it?' Whatever our point of entry, we each have the responsibility and opportunity to change the current reality.

I have argued that the current discourse in science and technology education has given a lot of power to very few people. Power refers to asymmetries between individuals or groups of individuals based on material, social, political or intellectual capital and access to structures. Power rewards and indulges some and sanctions others. It is therefore crucial that we understand these structures and become aware of the social and institutional arrangements that perpetuate the status quo. All voices must be heard in the discourse on science and technology education. Any monolithic voice should be drowned by a choir of 'others'. No special privileges should be granted and there can be no exceptions. Baumann (1993: 245) puts it as follows:

> What the post-modern mind is aware of is that there are problems in human and social life with no good solutions, twisted trajectories that cannot be straightened up, ambivalences that are more than linguistic blunders yelling to be corrected, doubts that cannot be legislated out of existence, moral agonies that no reason-dictated recipes can soothe, let alone cure. The post-modern mind does not expect any more to find the all-embracing, total and ultimate formula of life without ambiguity, risk, danger and error, and is deeply suspicious of any voice that promises otherwise. The post-modern mind is aware that each local, specialized and focused treatment, effective or not when measured by its ostensive target, spoils as much as, if not more than, it repairs. The post-modern mind is reconciled to the idea that the messiness of the human predicament is here to stay. This is the broadest of outlines, what can be called post-modern wisdom.

Such humility should guide us as science educators into a future where there is an absence of absolutism; where there are no assumed solutions, recipes and formulas; but where we all remain open to the possibility of learning from each other. Extending and enriching our understanding of the complexity of the issues that we face can be achieved only through authentic discourse.

## REFERENCES


Adas, M. 1989. *Machines as the Measure of Men: Science, Technology and Ideologies of Western Dominance.* Ithaca, NY: Cornell University Press

Baumann, Z. 1993. *Postmodern Ethics.* Oxford: Blackwell

Ernest, P. 1991. *The Philosophy of Mathematics Education.* London: Falmer Press

Fasheh, M. 1993. From a dogmatic, ready-answer approach of teaching mathematics towards a community-building, process-oriented approach. In Julie, C, Angelis, D & Davis, Z (eds). *Political Dimensions of Mathematics Education.* Cape Town: Maskew Miller Longman

Feyerabend, P. 1975. *Against Method.* London: Verso Books

Feyerabend, P. 1981. How to defend society against science. In *Scientific revolutions.* Ian Hacking (ed). Oxford University Press

Gibbs, W Wayt. 1995. Lost science in the third world. *Scientific American.* August

Katama, A. 1995. The African Forum for Children's Literacy in Science and Technology. A Profile of Activities

Kaufman, BA. 1975. Piaget, Marx, and the political ideology of schooling. *Curriculum Studies,* 10(1), pp 19–44

Longino, H. 1993. Subjects, power and knowledge. Description and prescription in feminist philosophies of science. In Alcoff, L & Potter, E (eds). *Feminist Epistemologies*, pp 101–20. New York: Routledge

Peat, F David. 1994. *Lighting the Seventh Fire.* New York: Carol Publishing

Reddy, V. 1995. Redress in science and mathematics education research in South Africa (unpublished paper)

Reiss, M. 1993. *Science Education for a Pluralist Society.* Buckingham, Pa: Open University Press

Ware, Sylvia A. 1992. *Secondary School Science in Developing Countries: Status and Issues.* The World Bank

Williams, P. 1961. *The Long Revolution.* Harmondsworth: Penguin Books

Zevenbergen, R. (in press). *Constructivism as a Liberal Bourgeois Discourse*


 

# 5

# Relevance in science and technology education

*Dr Marissa Rollnick, University of the Witwatersrand, Johannesburg, South Africa*

## ABSTRACT

The importance of the relevance of the science curriculum to successful learning in science and technology education is rarely questioned. This chapter does so. Was the curriculum in the past and is the curriculum in the present relevant to the needs of Africa? In addition, the author examines relevance to what and to whom in the future.

## RELEVANCE: ITS IMPORTANCE AND SOME QUESTIONS

### Introduction

The importance of relevance to learning is rarely questioned. Ausubel's now famous statement embodies it clearly: '... the most important single factor influencing learning is what the learner already knows.' However, education systems universally have neglected relevance, including the 19th-century educational system in the United Kingdom as satirized by Charles Dickens in *Hard Times* (Wilds and Lottich, 1971: 379).

> Sissy Jupe, girl no. 20, the daughter of a strolling circus actor, whose life, no small share of it, had been passed under the canvas; whose knowledge of horse, generic and specific, extends back as far as memory reaches; familiar with the form and food, the powers and habits and everything related to the horse; knowing it through several senses; Sissy Jupe has been asked to define horse. Bewildered by the striking want of resemblance between the horse of her conception and the prescribed formula that represents the animal in the books of the Home and Colonial/society, she dares not trust herself with the confusing description, and shrinks from it in silence and alarm. 'Girl no. 20 unable to define horse,' said Mr Gradgrind. Girl no. 20 is declared possessed of no facts in reference to one of the commonest of animals, and appeal is made to one red eyed Bitzer,

> who knows horse practically only as he has seen a picture of a horse or he has, perhaps, sometimes weathered the perils of a crowded street crossing. 'Bitzer,' said Thomas Gradgrind, 'your definition of a horse!!' 'Quadruped, omnivorous, forty teeth, namely twenty-four grinders, four eye teeth, and twelve incisors. Sheds coat in the spring; in marshy countries sheds hoof, too. Hoofs hard but requiring to be shod with iron. Age known by marks in the mouth.' Thus and much more, Bitzer. 'Now girl 20,' said Mr Gradgrind, 'you now know what a horse is.'

This quotation has many echoes in African classrooms, where science education is not only irrelevant, but its very irrelevance is considered a virtue. If it were relevant, it would not be considered education. Our colonial heritage has made us believe that education of necessity must be abstract and divorced from life. True academia is not vocational and vocational education is not academic, well satirized by Hooper (1971) in his description of a 'sabre toothed curriculum' in the educational system of a fictitious primitive society. Middleton (1988) enhances Hooper with his account of the indifferent history of secondary school vocational education.

## Defining relevance

Notions of relevance change over time and in different socioeconomic contexts. The relevance of schooling to the ordinary citizen in the African context has been primarily to obtain white-collar employment. Lewin (1992) comments, 'Life chances depend on educational qualifications in developing countries to a much greater extent than in industrialized countries.' Science is an important selection subject to enable students to progress higher in the educational system. However, the significance of science and technology education stretches beyond the narrow objective of producing white-collar workers. A consequence of irrelevance is that it stifles the economic and social potential for all strata of our society. Irrelevance affects quality of life and the ability of students to control their lives. If science and technology education is to have an impact on improving society, relevance becomes an essential ingredient of any meaningful programme. Making science relevant is part of making the subject accessible, which leads to motivation and achievement.

Just as there is more to changing curricula than changing the content, so there is more to relevance than providing relevant content. To help understand and strive for relevance in science and technology education, I ask these questions in this chapter:

- How do science and technology education policies in African countries address relevance and what mechanisms are in place to ensure their implementation?
- How has curriculum development promoted relevance in science and technology education, for whom is the curriculum intended, and what doors are opened by studying different curricula?
- What are the needs of those studying science and technology?
- Who defines science and technology?

## RELEVANCE AND WHAT ENSURES ITS IMPLEMENTATION

Government policy documents often do not explicitly state science and technology education policies. They frequently develop as a result of economic policies and can be either stated or unstated.

The structural adjustment programmes (SAPs) are the most influential economic policies of recent times in Africa. Lewin (1993) identifies the effect of SAPs on the development of science and technology education in African countries. He says, 'It is clear that hard policy choices may have to be made . . . .' Mbilinyi (1989) makes a scathing attack on a similar phrase, 'hard decisions on education policy should not be postponed', when she says:

> The level of arrogance in this report (the World Bank) is matched by its outright ignorance, probably a form of defensive ignorance — that is a political blindness towards aspects of reality that do not fit its particular set of preconceptions and goals.

The SAPs have affected teachers' earning power and resourcing of schools. Mbilinyi identifies some effects of SAPs as:

- Reduction of real wages for teachers, particularly at secondary and higher levels.
- Lengthening the working day and year for teachers, increasing class size, and consolidating of rural schools.
- Increasing costs of education for students and their parents.
- Reduction of state expenditure on education.
- Reduced enrolment at tertiary levels.
- Dependence on foreign experts.

Assertions are frequently made that university staff do not allocate much time and effort to direct service activities. In 1988 university staff in Tanzania could scarcely feed their families for three days on their monthly salaries. To live, they engaged in subsistence activities such as consultancies, poultry rearing or driving taxis. Such a climate precipitated by SAPs does not encourage reflective practice. Teachers working in four schools simultaneously are likely to use teacher-centred methods such as 'chalk and talk'.

Cross-country studies of cognitive achievement, such as the International Education Association (IEA) Science Study, test the science knowledge and understanding of students in different countries. Though Lewin (1993) concedes that such studies are flawed, he uses them to make comparisons which show that students from African countries perform poorly and score at the bottom of the table. However, those familiar with education in Africa can detect signs that the research design of the study may not take account of the African context. For example, the science tests were conducted on populations differentiated by age — a differentiation that makes sense in a country such as the UK where students progress from one year to the next. However, in Africa one commonly finds adolescents in lower primary classrooms and adults in secondary schools. Several science items favour urban over rural students in a continent that is mostly rural. The third IEA study, currently in progress (Robitaille, 1994), was even more unsuccessful than the second in securing the participation of African countries. Of the 50 participating countries, only two are from Africa — South Africa and Tunisia, neither typical of the African continent.

Achieving any meaningful change under the strain of SAPs is difficult. Lange (1995) describes a subtle attempt to bring systemic change to education in Zanzibar. Science camps for students serve as a microcosm of the educational system where ministry officials can try new approaches to change in a supportive, non-threatening environment. The next step for such officials is to transfer their vision of the possibilities for change to the larger educational system.

Different stakeholders define relevance in different ways. This results in different programmes and policies, depending on which is the dominant group. The aims of most science curricula state they want children to think scientifically, but rarely realize this in practice that is generally determined by examinations. Students and parents regard entry to the job market as the most important reason for schooling. Since passing examinations is a prerequisite for a job, enabling their students to do so becomes the aim of most teachers. Thus, the need to understand science for relevance, scientific literacy and preparing the scientists of tomorrow becomes lost.

Lewin (1992) writes of the conflict between job providers and job seekers. University subject specialists are often blamed for their influence on the content of science education courses. They usually receive their postgraduate training abroad and return espousing the philosophy of the country in which they received their training. However, lacking the support given postdoctoral research in more developed frameworks, and faced with large teaching and administrative responsibilities, their research suffers. Those who enter other sectors find that their work is largely routine (Lewin, 1992). Having lacked exposure during their training to research using appropriate technologies, research scientists' training is inappropriate.

Ogunniyi (1986) criticizes the esoteric science programmes offered by many universities in Africa and the isolation of African scientists from the debate on curriculum development. The opposite is the case in South Africa, where universities dictate subject content, resulting in teaching of decontextualized science.

## SOME ATTEMPTS TO DEVELOP A RELEVANT CURRICULUM

In most countries in Africa, science was introduced as a school subject only after independence. Before that, education was largely restricted to the provision of primary schooling in which science did not play a significant role. Science taught to the minority who attended secondary school was a reflection of that in the schools of the colonial master (Lewin, 1992). Students in the British colonies sat examinations set by overseas examining boards that still examine students in some southern African countries. Despite attempts to modify such examinations, it is difficult to set questions that will be simultaneously relevant in Singapore, the Bahamas and Swaziland, and a brief survey of past papers set by the Cambridge Overseas School Certificate confirms this. All the papers for the chemistry section of combined science (IMSTIP, 1989) are entirely stripped of context. Only references to a lollipop, grape juice and red ink come from outside the sanitized world of the chemistry laboratory. Failure to relate to the terms neither assists nor interferes with the candidates' ability to answer the question.

Ogunniyi (1986) quotes Yoloye and Bajah (1981) in his description of curriculum innovations in Africa when he says that curriculum change is perhaps the most remarkable change that has occurred in African countries since independence. Influenced by the post-Sputnik wave of curriculum development in the industrialized North, changes in African secondary schools were mostly adaptations of overseas curricula (Lewin, 1992). Many of these courses aimed to produce scientists (Buttle, 1975), and were designed for the top 30 % of students in the host country (Lewin, 1992). Africa rejected the vocational or technology-oriented courses developed in those countries around the same period, since Africans regard schooling as preparation for white-collar employment. Educators and students regarded 'relevance' as vocational and thus a vice rather than a virtue.

The African Primary Science Programme (APSP) spearheaded curriculum change at primary levels in Africa. Unlike materials produced for secondary programmes of the time, APSP materials were developed in Africa.

By the late 1980s most countries in Africa had established curriculum development centres (Lewin, 1993). Evaluation studies show that despite efforts by these centres and international agencies to change teachers' pedagogy, few have done so. More recent developments such as BOTSCI — a junior science programme for Botswana — have been achieved with less outside assistance (Nganunu, 1988). ZIMSCI, a science programme in Zimbabwe, inspired BOTSCI and used a low-tech, kit approach (Kahn & Rollnick, 1993). With BOTSCI, the country made a policy decision to teach a 'science for citizens' course at the junior secondary level, and to prepare future scientists and technologists at later stages of education. Science and Technology in Society (STS), a British course, influences BOTSCI, though it was impossible to adapt the highly contextualized STS materials. Where environmentally contextualized materials of this nature are developed, relevance means more than changing content or methods of teaching.

A weakness of many of these developments was a failure to involve personnel at all levels of the educational system (Lewin, 1992), particularly those involved with examinations. A further difficulty was that rapid expansion of the education system and a shortage of foreign exchange strained the ability of African countries to provide the necessary support for the developments. Problems experienced by ZIMSCI exemplify this situation (Kahn & Rollnick, 1993).

An exciting recent event in African curriculum development took place in Harare, Zimbabwe in January 1991 (Whittle et al, 1993). Described as a 'generator' of ideas rather than a conference, the event aimed to:

- gather people of proven creativity and enterprise;
- involve officials and policy makers;
- benefit some nonparticipant Zimbabwean teachers;

- facilitate the production of products that demonstrated innovative approaches to science and technology education in Africa;
- distribute the product as widely as possible within Africa; and
- encourage the use and adaptation of the products.

An intensive week of writing, trialing and critique of materials followed an introductory week of exposure to new ideas by a team from the United Kingdom (UK) which had been involved in the design of STS. The product is an exciting collection of multimedia materials relevant to Africa in terms of content, equipment and teaching methods that exemplify that methods and content must be relevant.

Lewin (1992) identifies five factors that militate against the adoption of relevant curricula by countries in Africa:

1. Pressure for continuity between stages, even if only a minority continue with schooling.
2. Difficulty in finding teachers with the necessary qualifications and experience to make science teaching relevant.
3. The fact that science is often used to discriminate between students.
4. Those involved in defining the subject are themselves successful products of the established system.
5. The science of everyday life has a low status.

A possible way forward is to make 'science of everyday life' a subject for all students, as is the case with junior secondary science in Botswana (Nganunu, 1992). However, concentrating the curriculum on everyday issues could limit learners' horizons.

Lubben et al (1995) describe interesting research that promoted relevant curricula in Swaziland. The action research model facilitated the production of learning materials designed to be contextualized within the realities of Swaziland, as well as being applicable and open to investigation. The project included teachers in the design and trialing of materials and focused heavily on teacher development in its implementation. Research showed that, while students showed few signs of improved cognitive improvement, there were noticeable gains in the affective domain for both boys and girls. Course materials were particularly effective in maintaining girls' interest in traditionally male topics, such as electric circuits. Improvement in the affective domain is probably more important than cognitive gains, since they may lead to increased student interest and effort. The revised project materials are similar to conventional integrated science materials, perhaps revealing that the project teachers were not comfortable with a radical departure from what they know.

A Nigerian study (Otuka, 1993) of teachers' views on the effectiveness of various textbook attributes revealed that important factors are: use of local examples; local alternatives to scientific terminology; identification of locally practised science and technology; use of local languages; reflection of local climatic and environmental factors; identification and demystification of taboos and superstitious beliefs;

consideration of local tools, occupations and agricultural products; reflection of variations of local buildings; and use of cooperative problem-solving techniques.

However, despite an interest by researchers, science educators and donors in promoting relevance in curricula, policy and examinations have yet to institutionalize this interest.

## MAKING SCIENCE RELEVANT

### Science for all

Science and technology learning for all is an equity rather than a relevance issue. However, once it has been raised, the challenge of providing students with a relevant experience becomes of prime importance. 'Science for all' implies relevance and comprises the knowledge and skills needed to empower students to control their lives at an individual and a societal level.

At the individual level 'science for all' could mean understanding waterborne diseases, how to purify water, and the basic principles that underlie these issues. At the societal level it could mean having the confidence, skills and knowledge needed to challenge the management of a factory that pollutes a river.

A common misconception of 'science for all' is that it is inferior and therefore not suitable for able students. However, able students and especially those who will become scientists and technologists need to understand societal issues. Designers of course materials for 'science for all' must consider whose science and what science, since society must also respect local scientists and technologists, such as traditional healers.

Issues of gender regarding 'science for all' are discussed by Vijay Reddy in chapter 6.

### Teachers

African societies judge teachers by their ability to help students to pass examinations, since passing or failing can mean the difference between white-collar employment or sweeping the streets (Lewin, 1992: 105).

Though many public examinations may test higher cognitive skills, they remain decontextualized and content-driven, and this defines relevance for teachers. Anything that forms part of the examination syllabus is relevant and teaching methods other than drill are rarely seen.

Teachers in Africa are underpaid and frequently do several jobs to feed their families. Considerations other than financial must motivate them to continue teaching. NEPI (1992), for instance, reports that what teachers value most about in-service courses is the collegial contact.

Appropriate teacher development is crucial. Studies by Kelly and Rollnick (1996), Kannieappan (1996: 30), Wuyep and Turner (1994) in Nigeria, and Klindt (1994) in Lesotho established the importance of relevant content and teaching methodologies in both pre-service and in-service science and technology teacher development programmes.

 

### Students

Unlike in the rest of Africa where students view school as a means to a job, in South Africa students' demands for relevant education spearheaded political change. The 1976 disturbances over the medium of instruction soon expanded into a demand for people's education and a concern for what schools taught. Kahn and Rollnick (1993) speculate how this movement applied to science education. The decontextualization of science teaching led to a lack of the ideological distortion that occurred with other subjects during the apartheid regime. However, a high failure rate led to a fear of and aversion to science, as exemplified in the apocryphal tale of the Soweto student who stated there is no place for mathematics in a people's education since it is divisive, and students who did not understand it would feel inadequate. However, demands made by students in the name of people's education related to content, not to teaching methods.

Educational unrest in South Africa resulted in the formation of the National Education and Training Forum (NETF). This body represented important stakeholders in the education process, such as teachers' unions, government, and school and university student bodies. A short-term syllabus initiated by the government in 1994 was unique in that it included secondary and university students in the drafting process (Rollnick, 1994). Student participation is important in shaping the science curriculum.

## WHO DEFINES SCIENCE?

Horton (1971) wrote one of the earliest papers that addresses the issue of control. While he alleged that there are similarities between Western and African views, he noted important differences regarding openness, especially regarding perceptions of alternatives and possible threats to established bodies of knowledge that influence students' understanding of science. Adu-Ampoma (1975) cites unquestioned belief in authority as an aspect of traditional African thought that limits students' learning. The ideology of science as a fixed body of knowledge with 'correct' answers pervaded science curricula in apartheid South Africa, where a white minority defined science and science teaching (NEPI, 1992). Studies abound by Western and African researchers on the influence of traditional modes of African thought on science learning (Kay, 1975; Sawyer, 1979; Ingle & Turner, 1981; and Ogunniyi, 1987). Ogawa (1986) suggests a model to explain the conflicts between traditional and scientific thinking and suggests that a process of simultaneously exploring Western and traditional African ideas would result in a deeper understanding of both. Rollnick (1988) sums up the contradiction:

> The student in Africa has one name which is used at school and another one which is used at home. There is one type of acceptable behaviour at school and one at home. There is one type of dress for school and one for home. There is a language for school and a language for home. Because of this, the student, too, becomes two people. Why not two concepts of science?

 

When one looks at the relationship between traditional thinking and school, one ventures into the fraught area of culture. A series of articles in the *Journal of Cross-cultural Psychology* in 1984 reflects this (Rohner, 1984; Jahoda, 1984; and Segall, 1984). Rohner rejects behaviourist definitions of culture, preferring to see it as: 'The totality of equivalent and complementary learned meanings maintained by a human population or by an identifiable segment of the population and transmitted form one generation to the next'. Jahoda criticizes this view saying that it draws the line between ideas and behaviour too sharply. Segal, on the other hand, feels that it is pointless to search for a definition as doing so does not advance the study of cross cultural psychology. To avoid this problem, Toulmin (1972) uses the term 'conceptual ecology'. A more productive term may be 'intellectual environment' (Hewson & Hamlyn, 1983).

Many of the above studies argue that the design of curricula takes account of 'culture' or 'intellectual environment', thus negating the positivist view of science as a 'value free' subject. Teaching methodology is related to the adaptation of curricula. Some argue that the student-centred methodologies proposed in modern curricula are contrary to 'African culture'. However, there is a growing body of opinion even in Western teaching situations that teacher-centred lessons are often effective. Wildy and Wallace (1995), after conducting a study of a teacher who used whole-class teaching almost entirely, argued for a '... broader view of good science teaching than that proposed by the literature, one that takes into account teachers' and students' understanding of science in relation to their social and cultural contexts'.

## SOME SUGGESTIONS ON THE PROMOTION OF RELEVANCE

Confronting the difficulties caused by structural adjustment programmes requires action at a broader level than science education. African countries need to find ways forward that will allow them to break from the World Bank and the International Monetary Fund (IMF). However, science and technology education may find creative ways to progress. One possibility is work along the lines reported by Lange on the science camps in Zanzibar, where civil servants are freed temporarily to exercise their creativity and see beyond the rules that bind them. Another option is to find ways to help overloaded and underpaid teachers to teach science in a meaningful way. Often they stay in teaching merely because it is more stimulating than some of their more lucrative income-generating activities. A fundamental shift in policy that could make a difference to what happens in classrooms would be to allow those responsible for developing the curriculum to have more control over the examination process even if the dominance of examinations cannot be diminished.

The development of STS-type programmes is promising even though the concept of STS is an import. The nature of STS programmes is that they must be contextualized and developed locally. They are thus a useful vehicle for developing self-reliance and ensuring that course resources are not expensive.

 

The tyranny of examinations will not be banished easily as too much is at stake for successful candidates. A logical policy option would be to make the system promote relevance by redesigning examinations to do so, thus placing it on the agenda of teachers, students and parents.

## REFERENCES

Adu-Ampoma, SM. 1975. Myth and superstition: The African child's background. *Science Teacher,* 18(3–4), p 21

Buttle, J. 1975. Chemistry and the curriculum. In Daniels, DJ (ed). *New Movements in the Study and Teaching of Chemistry.* London: Temple Smith

Hewson, MG & Hamlyn, A. 1983. The influence of intellectual environment on conceptions of heat. Paper presented at the annual meeting of the American Educational Research Association

Hooper R (ed). 1971. *The Curriculum: Context, Design and Development.* Edinburgh: Oliver and Boyd

Horton R. 1971. African traditional thought and Western science. In Young, MFD (ed). *Knowledge and Control.* Milton Keynes: Open University Press

IMSTIP. 1989. *Science Chemistry Past Examination Papers (with Model Answers) 1984–1988.* Inservice Maths Science Improvement Programme, Swaziland

Ingle, R & Turner, A. 1981. Science curricula as cultural misfits. *European Journal of Science Education,* 3(4), pp 357–71

Jahoda, G. 1984. Do we need a concept of culture? *Journal of Cross-cultural Psychology,* 15(2), pp 139–51

Kahn, M & Rollnick, M. 1993. Science education in the new South Africa: Reflections and Visions. *International Journal of Science Education,* 15(3), pp 262–72

Kannieappan, A. 1996. The status of physical science classrooms: A case-study of ex-House of Delegates schools. Unpublished MEd dissertation. Durban: University of Durban-Westville

Kay, S. 1975. Curriculum innovation and traditional culture: A case study of Kenya. *Comparative Education,* 11, pp 183–91

Kelly, G & Rollnick, M. 1996. Higher Diploma in Education students' knowledge of concepts related to chemical bonding. Paper presented at the annual meeting of the Southern African Association for Research in Mathematics and Science Education. Pietersburg, South Africa

Klindt, P. 1994. The Lesotho induction programme (1994). *Science Education International,* 5(4), pp 13–15

Lange, R. 1995. How do systems of education change and how shall we change our own? Some modest generalisations. Links Proceedings of the 16th National Convention of Natural Science and Mathematics Education Associations of South Africa, Johannesburg 10–16 July, pp 125–32

 

Lewin K. 1992. *Science Education in Developing Countries: Issues and Perspectives for Planners.* Paris: International Institute for Educational Planning

Lewin, K. 1993. Planning policy on science education in developing countries. *International Journal of Science Education,* 15(1), pp 1–15

Lubben et al. 1995. In-service support for a technological approach to science education. Overseas Development Administration Education Paper Serial no 16. London: ODA

Mbilinyi, M. 1989. Crisis of education and research in the 1980s — Challenges for the future. Proceedings of the Boleswa Symposium on Educational Research, Gaborone

Middleton, J. 1988. Changing patterns in vocational education. World Bank policy planning and research paper. Working Paper WPS 26 (mimeo). Washington: World Bank

NEPI. 1992. Science Curriculum Report. Johannesburg: National Education Policy Initiative, mimeo

Nganunu, M. 1988. An attempt to write a Science Curriculum with Social Relevance for Botswana. *Journal of Science Education* 10(4), 441, 448

Nganunu, M. 1992. Inclusion of indigenous technology in school science curricula — a solution for Africa? In Yager, R (ed). *The Status of Science Technology Society Efforts Around the World.* ICASE

Ogawa, M. 1986. Towards a new rationale for science education in a non-Western society. *European Journal of Science Education,* 8(2) pp 113–19

Ogunniyi, MB. 1986. Two decades of science education in Africa. *Science Education,* 70(2), pp 111–22

Ogunniyi, MB. 1987. Conceptions of traditional cosmological ideas among literate and non-literate Nigerians. *Journal of Research in Science Teaching,* 24(2), pp 107–17

Otuka, JOE. 1993. Teachers' views on effective primary science in Nigerian schools. *Science Education International,* 4(1), pp 23–5

Robitaille, DF. 1994. The Third International Mathematics and Science Study: An overview. *Science Education International*, 5(4), December pp 27–34

Rohner, RP. 1984. Towards a conception of culture for cross-cultural psychology. *Journal of Cross-cultural Psychology,* 15(2), pp 111–38

Rollnick, M. 1994. Assessment aspects of the short-term curriculum change process. Paper presented at the CASME conference on assessment. Durban, November 1994

Rollnick, M. 1988. Mother-tongue instruction, intellectual environment and conceptual change strategies in the learning of science concepts in Swaziland. Unpublished PhD thesis. Johannesburg: University of the Witwatersrand

Sawyer, ES. 1979. The role of traditional beliefs in the teaching and learning of science in Sierra Leone. *Science Education*, 22(3/4), pp 3–42

Segall, MH. 1984. More than we need to know about culture but were afraid to ask. *Journal of Cross-cultural Psychology,* 15(2), pp 153–62

 

Toulmin, S. 1972. *Human Understanding vol 1: The Collective Use and Evolution of Human Concepts.* Princeton: University of Princeton Press

Whittle, P, Gray, B, Hodzi, R & Manana, L. 1993. *Innovative Ideas and Techniques for Science Educators in Africa.* Zomba: ICSU

Wilds, EH & Lottich, KV. 1971. *Foundations of Modern Education.* New York: Holt Rinehart and Winston, p 379

Wildy, H & Wallace, J. 1995. Understanding teaching or teaching for understanding: Alternative frameworks for science classrooms. *Journal of Research in Science Teaching,* 32(2), pp 143–56

Wuyep, E & Turner, M. 1994. Integrated science teacher education in Nigeria: How effective is it? *Science Education International,* 5(4), pp 13–15

Yoloye, EA & Bajah, ST. 1981. *Science Education for Africa, vol I: A Report of Twenty Years of Science Education in Africa.* SEPA

# 6

## Relevance and the promotion of equity

*Vijay Reddy, University of Durban-Westville, Durban, South Africa*

### ABSTRACT

Historically, the participation of girls in science and technology education has been poor. In some parts of Africa certain racial groups and nomadic tribes were discriminated against, resulting in their poor participation in science and technology education. With the advent of 'science for all', equity in science and technology education has become an imperative. This chapter focuses on the challenges of access, redress, equity, and quality in science and technology education. It analyses past and present trends and proposes future directions with regard to these challenges.

### INTRODUCTION

Ogunniyi (1995), writing in *Science Education*, sets a backdrop for discussing science education in Africa:

> Since their independence in the late 1950s and 1960s, most African states have become acutely aware of the importance of science education as a means to scientific and technological development ... Within the continent, the two major declarations adopted by African heads of state and government, the Lagos Plan of Action (1980) and the African Priority Programme for Economic Recovery (1986), have both called for sustainable development based on self reliance in science and technology applications. A dominant theme has been that without a sound science education programme a country cannot achieve any breakthrough in its economic development (OAU, 1981).

Ogunniyi goes on to say '. . . the state of science education in Africa today is far worse than was reported earlier' (Ogunniyi, 1986).

Various papers and reports have documented the state of education and science education. It is recognized that, since the 1970s, education systems in sub-Saharan Africa have brought a measure of basic literacy and numeracy to more than half the population. However, they have failed to produce sufficient numbers of technical and managerial workers with the skills to meet the needs of modernizing the economy.

Science educators have identified problems such as lack of resources for teaching science, inadequate laboratory facilities, too few qualified science teachers, and large classes. There are further disparities within this context of impoverishment. These are disparities of class, gender, race, location and poverty. Often the issue of equity affects several overlapping disadvantaged groups, such as rural poor girls, making them groups that are the most disadvantaged.

I shall describe briefly the state of education and science education in some countries, and then discuss a framework to achieve equity in science and technology education.

## THE STATE OF EDUCATION AND SCIENCE EDUCATION IN AFRICA

### Participation in schooling

All governments in Africa are committed to universal access to primary education. However the gross enrolment rates (GERs) vary. Few countries have reached 100 % and where overall participation is low, gender gaps are wider. In countries where there was exclusion by race, it has affected participation patterns. In all countries children of poor families have low school enrolment and high dropout rates. Proximity to schools also affects enrolment. Anderson (1988) reports that the International Council of Education Development estimates that fewer than 50 % of rural children in most countries and as few as 10 % in some countries complete four or more grades in school. In Sudan, for example, 80 % of urban but only 20 % of rural children go to school (*West Africa Weekly Magazine,* 1989).

In 1990, in Africa, girls made up 45 % of primary and 40 % of the secondary school population (Odaga & Henneveld, 1995). Africa-wide, enrolment rates (percentage of the age group) of 6–11 year olds is 69 % for boys and 57 % for girls (Lockheed & Verspoor, 1991), though in Botswana, Lesotho, South Africa and Mauritius there are more females than males enrolled in the education system. Participation patterns at the secondary level show that the gap between boys and girls widens further. For example, Zambia's secondary school population in 1994 was made up of 62 % boys and 38 % girls (Nair & Tindi, 1995).

In Zambia, a lack of school space denies access to formal education to 45 % of the seven-year-olds (Nair & Tindi, 1995). In Nigeria, about 23 % of primary school pupils are not in schools (Ivowi, 1995). In South Africa the primary school enrolment rate is about 71 %. Botswana has achieved almost universal primary education.

GERs at secondary school levels vary from about 50 % in countries such as Zimbabwe and South Africa to less than 10 % in Malawi and Tanzania (Lewin, 1996).

### Participation in science and technology education in schools

All pupils in primary schools study science, where it is called names such as environmental science, integrated science and general science. Participation in science at the secondary school level is a function of the secondary GER and the proportion of those who study science at secondary school. There is low participation in physical sciences at high school as well as a gender and race gap in enrolment patterns in biology, physics and chemistry, with the greatest difference in the physical sciences.

In most countries students in grades 10–12 are required to take some science. Most students study biology, with a few taking chemistry, physics or physical science (Lewin, 1996). For example, in Kenya in 1994, of all the candidates registered in the senior examinations, 22 % studied physics, 42 % chemistry and 58 % physical science (Wasanga, 1995). In Nigeria in the senior secondary school about 93 % studied biology, about 30 % chemistry and 16 % physics (Okebukola, 1995). In South Africa in 1990, of all the standard 10 pupils, 36 % studied mathematics, 22 % physical science and 76 % biology (FRD, 1993).

Within the low participation in science there are further disparities by race and sex. In South Africa, in 1990, at standard 10, 47 % of white pupils and only about 15 % of African pupils took physical science as a subject (FRD, 1993). In Zambia the ratio of boys to girls studying chemistry and physics in grade 12 in 1994 was about 86 % male and 14 % female (Chibesakunda, 1995). In Zimbabwe girls constitute about 20 % of the total number of A-level students enrolled in science subjects (Zimbabwe Ministry of Education, 1995).

## Performance in science in schools

Performance in science subjects is generally poor. In 1993, performance in English, mathematics and science in the Kenyan Certificate of Primary Education (KCPE) shows the following pattern of A grades awarded in the examination (Wasanga, 1995).

| | **English** | **Maths** | **Science** |
|---|---|---|---|
| Boys | 5 144 (51 %) | 11 082 (75 %) | 3 609 (88 %) |
| Girls | 4 923 (49 %) | 3 639 (25 %) | 489 (12 %) |
| Total | 10 067 (100 %) | 14 721 (100 %) | 4 098 (100 %) |

There are fewer A grades in science than in English or mathematics; there are also gender disparities in mathematics and science. Performance in science at the Kenyan Senior Certificate Examinations in 1994 was also poor. Of those who wrote the examination, the percentages of candidates being awarded any grade higher than D+ were: 14 % in mathematics; 64 % in biology; 48 % in physics and chemistry, and 25 % in the physical sciences (Wasanga, 1995).

In Zambia, a baseline study for the grade 9 project, 'Action to Improve English, Mathematics, and Science' (1994), showed that girls achieved a pass rate of 31 %, and boys of 57 % (Nair & Tindi, 1995). In Swaziland, in the mathematics/physical science combination, three times as many boys as girls obtained a credit pass (UNECA Report, 1990).

 

In South Africa, student performance in mathematics and physical science shows an alarming picture by race. The matriculation pass rates in 1990 for white, coloured and Indian pupils were relatively high, with 95 % passing physical science, 91 % passing mathematics and 88 % passing biology. For black pupils, the results were dismal, with only 15 % passing mathematics, 44 % passing physical science and 29 % passing biology (FRD, 1993).

### Performance in tertiary institutions

Lower enrolments of disadvantaged groups in sub-Saharan Africa are most pronounced in higher education, largely as a consequence of the inequities experienced at the primary and secondary levels. In 1990 girls made up 31 % of the tertiary population, with limited female representation in science, mathematics and technical courses (Odaga & Henneveld, 1995).

Only about 38 % of the small population which attends secondary school in Zambia proceed to some form of tertiary education (Nair & Tindi, 1995). Graduation figures in 1988 show that eight women graduated in natural sciences. None graduated in mining or engineering (Nair & Tindi, 1995).

In South Africa, one of the consequences of apartheid policies is poor school performance by black students. The number of degrees, diplomas and certificates awarded by universities reflects this. In 1991, of the 3 341 degrees in natural sciences and mathematics awarded by South African universities, 370 were awarded to Africans and 2 563 were awarded to whites (FRD, 1993).

### Participation in the workplace

The ILO estimates that in 1990 women formed 38 % (73 million) of the total labour force in sub-Saharan Africa, of which 76 % worked in agriculture, 17 % in the informal sector, and 5 % in the modern sector. Within the modern sector, women are employed mainly in the civil service, usually at the lower grades (Odaga & Henneveld, 1995). There are few women managers and their representation in central government and political parties remains weak.

## SOME REASONS FOR DISPARITIES

A reason for low GERs in schooling, low participation in science and poor performance in science education is an interaction between supply, demand and the learning process. Supply refers to the availability and quality of school facilities, materials and teachers. Decisions made by parents, based on opportunity, costs of schooling, religion and culture, create the demand. The learning process involves the experiences that children have in school that are linked to the curriculum (Lockheed & Verspoor, 1991). Disparities between groups arise for different reasons.

One reason for the lower participation of girls is a lack of demand because of family and societal views about schooling for girls. Furthermore, curriculum inadequacies and different treatment in the classroom of female students by both male and female teachers affect performance.

 

Schools are far apart in rural areas with a low population density. Inhabitants are generally poor and cannot absorb the extra costs of schooling. For children living far from school there are transport costs (if transport is available), and time spent walking to and from school reduces time for household maintenance and production chores, especially for girls. In poor families — particularly in rural areas — such child labour is often critical to family survival. Ndunda and Munby (1991) report that in Kenya 'traditional son preference still influences rural parents, who remain unwilling to invest in their daughters' education because the investment is considered wasteful or frivolous'. Such factors affect girls more than boys in rural areas, so gender differences are more acute when desegregated by urban–rural residence.

Many rural schools offer only three or four grades and lack resources such as teachers, materials, facilities and equipment. They often have more than one grade level per class. Teachers either treat the whole class as a single grade level or, if there are two grades per class, each grade level gets half the attention. Also, the language of instruction may not be that of the local population, and often the curricula are taught in a national, urban language that is not used in rural areas.

During the apartheid era, South African education was based on a philosophy of 'what is the point of teaching a Bantu child mathematics when he cannot use it in practice' (Hendrik Verwoerd: Hansard). This philosophy led to an education for blacks that was characterized by underspending, a lack of facilities, overcrowded classrooms, and unqualified or poorly qualified teachers.

## EQUITY IN SCIENCE AND TECHNOLOGY EDUCATION

Achieving equity in education is important because of its relationship to economic development and social justice. Many countries with successful recent histories of economic development have invested heavily in human resource development at primary and later at secondary levels, achieving approximately universal levels of enrolment. These countries have also stressed science, technology and mathematics during the period when economic growth was most rapid. To ensure economic growth, one of the necessary but not sufficient conditions is that all the population be educated (Lewin, 1992). In a democratic country, all groups should receive a quality education.

Harding (1992) and Erinosho (1994) list reasons why science and technology education should involve girls and women. The reasons would apply to all disadvantaged groups and are compelling:

1. Equality of opportunity is necessary so both sexes can be part of mainstream development.
2. Equity is important for technological and socioeconomic development.
3. There is a need for more female scientists in decision-making positions to enable them to control the direction of technological research and promote policies that favour females.
4. Science is exciting and its study promotes intellectual understanding, exploration and mastery of the environment.

5. If women suffer discrimination in science and technology, a lack of appropriate qualifications will limit their financial rewards and bar them from interesting work.
6. Being excluded from science would lead to a sense of alienation among women and, with modern life becoming increasingly dependent on science and technology, such alienation would not be healthy.

## ISSUES TO CONSIDER WHEN PLANNING FOR EQUITY

Policy is a blunt instrument to produce intended educational change. Reasons why change has not yet been effected include: (1) a shortage of well-trained and motivated teachers; (2) a failure to implement planned curricula because of a lack of resources; (3) a failure to consider prevailing national conditions; (4) not involving teachers in policy formulation; and (5) a lack of planning and coordination between those institutions concerned with provision of science education (Caillods, Gottelmann-Duret & Lewin, 1995). Achieving equity will require more than a change of educational policy. Educational planning, programming, management, implementation, monitoring and evaluation must all have an equity perspective. To achieve this will require a strong political will.

Considering the depressed economies of many African countries, we must think of ways to effect change within present government budgetary constraints, though some programmes may require donor support. When resources are limited, there is always a policy dilemma between providing education to all and ensuring an equitable distribution of resources. For example, a study by Obura in 1991 (quoted in Caillods et al, 1995) illustrated laboratory costs in different locations of Kenya. A low-cost laboratory near Nairobi costs $20 000; $32 000 in a rural area near a tarred road; and $40 000 in a rural area off the tarred road.

We cannot look at issues of science education and disparities in isolation. They must be considered within the context of the educational system which in turn must be within the context of sociopolitical and economic systems. Education ministries alone cannot achieve equity. It requires commitment from all ministries as well as changed attitudes within society and the workplace. Thus, changing classroom practices only cannot achieve equity. We must examine school, societal and family practices; perceptions of schooling; political and institutional factors; individual factors; workplace opportunities; and the economic status of the family.

Improving educational opportunities for all children is a key factor in promoting the enrolment of girls. We can increase the demand for schooling by changing parents' and communities' perceptions and by demonstrating its usefulness by ensuring that the workplace can absorb school graduates. There should be visible results to encourage others, and we must ensure that children have a good learning experience so they are more likely to want to remain in school.

For me, achieving equity is a process with both a short-term and a long-term goal. The long-term goal is to eliminate disparities by ensuring equitable participation and performance. Attempts to change participation rates by changing only school/classroom practice have not been effective. Research shows that sociocultural practices prevent girls from participating, continuing, and performing well in school. At the risk of being called elitist, I therefore suggest a programme of targeted intervention as a short-term measure to achieve the long-term target of universal primary and secondary education for girls, and to increase their performance. A cadre of highly qualified and well-placed women would have the effect of: (1) changing societal perceptions about educated women; (2) creating positive role models; and (3) having a critical mass of women in organizations to ensure that their practices change.

Other equity issues in science and technology education include:

1. The efficacy of single-sex schools and segregated classes in promoting the participation and performance of girls. Studies conducted in Germany, and quoted in the *CASTME Journal*, show that cooperative learning in girls-only science classes in coeducational schools promotes the most improved performance in science learning.
2. Anecdotal evidence suggesting that women teachers encourage participation and performance by female students.
3. Evidence which suggests that male and female teachers believe that male students are academically superior to female students. Other studies (Zonneveld, Taole, Nkhwalume & Letsic, 1993) show that many classroom behavior patterns of teachers favour boys and affect both the performance and attitudes of girls in science and mathematics.
4. Students from poor households drop out of school to engage in income-generating activities or household maintenance tasks. Perhaps we should plan for school *and* work, rather than school *or* work (Odaga & Henneveld, 1995).
5. Ignorance on the part of parents and the community of the value of schooling, the nature and role of science and technology, and about science-and-technology-related professions.
6. The inability of poor households to afford schooling.
7. Many studies indicate that students experience difficulties when they learn science through a second or third language. Rural students are often taught using a national language that is the language of the urban population. This exacerbates their learning difficulties.
8. Teaching in multigrade classrooms generally leads to coverage of only a fraction of the syllabus and to insufficient practical work being offered students.
9. A lack of curricula that are appropriate to teaching relevant 'science for all'. Marissa Rollnick deals with this issue in chapter 5 above.

Considering the current economic constraints in many African countries, there is a dilemma between providing education to all and promoting equity. With respect to science and technology education, it might be wise both to change the state of science education and to engage in intervention programmes targeted at specific groups. However, targeted interventions are costly and frequently elites resist such programmes, fearing the implications of a redistribution of resources.

## RECOMMENDED SCHOOL STRATEGIES FOR PROMOTING EQUITY

Considering the issues discussed, I make the following, tentative recommendations:

1. Studies similar to those conducted in Germany and quoted in *CASTME Journal* 15(2) should be replicated in Africa.
2. Efforts should be made to increase the number of female science teachers.
3. Female science teachers should be provided with incentives to teach in rural schools.
4. All teacher education programmes should incorporate activities that make teachers aware of how certain practices disadvantage girls.
5. Targeted intervention programmes should be established that develop cadres of elites and provide employment opportunities. Such programmes would require commitment by government, NGOs and the private sector.
6. School timetables must accommodate working children from poor families who have to generate an income. A flexible school timetable could offer evening classes so that children can attend school after they have completed their household and other chores.
7. Programmes should be mounted through school boards and the mass media to educate parents and the community on the value of schooling, the nature and role of science and technology, and about science-and-technology-related professions.
8. Scholarship programmes should be set up to encourage poor pupils. To ensure that targeted groups continue to study at universities there need to be bursary and scholarship programmes. Donor communities should support such schemes.
9. A centre of excellence with a strong emphasis on science and technology should be established in a rural area, with a quota for girls. This initiative will require strong support from outside agencies.
10. Science teacher training programmes and science curriculum materials should incorporate language training.
11. All teacher training programmes and curriculum materials should contain components that help science and technology teachers work in multigrade classrooms.
12. Outstanding girls in the primary and secondary school system should be supported to move into tertiary education and should subsequently be provided with high-level, visible jobs in the government and the private and public sectors.

## CONCLUDING REMARKS

When addressing issues of equity, we must not treat disadvantaged groups as problems. Achieving equity involves the interaction of a number of issues. Ensuring equity rather than merely providing equal opportunity necessitates an analysis of class structures followed by praxis, not of inconsequential tampering with educational systems. Therefore a holistic approach that considers the individual, society, family, learning institution and the workplace must be supported.

## REFERENCES

Anderson, MB. 1988. Improving access to schooling in the third world: An overview. Bridges. Research Report Series, Issue 1. Cambridge, Ma: Harvard University

Caillods, F, Gottelmann-Duret, G & Lewin, K (forthcoming). *Planning Secondary Science Education.* UNESCO

*CASTME Journal*, 15(2), 1995, p 15: Girls learn better on their own

Chibesakunda, GA. 1995. Science education in Zambia. Paper submitted to *Planning Science Education at the Secondary Level.* Johannesburg: IIEP and CEPD

Erinosho, SY. 1994. *Girls and Science Education in Nigeria.* Anglo International Publishing, Nigeria

FRD (Foundation for Research and Development). 1993. *South African Science and Technology Indicators.* Pretoria: FRD

*Hansard.* 1954. Parliamentary Record South Africa

Harding, J. 1992. *Breaking the Barrier: Girls in Science Education.* Paris: IIEP

Ivowi, UMO. 1995. Science education at secondary level in Nigeria. Submitted to *Planning Science Education at the Secondary Level.* Johannesburg: IIEP and CEPD

Lewin, KM. 1992. *Science Education in Developing Countries: Issues and Perspectives for Planners.* Paris: IIEP

Lewin, KM. 1996. *Planning Secondary Science Education: Progress and Prospects in the African Region.* Paris: UNESCO

Lockheed, ME & Verspoor, AM. 1991. *Improving Primary Education in Developing Countries.* Oxford University Press. Washington DC: World Bank

Nair, A. & Tindi, E. 1995. The status of science education in secondary schools in Zambia. Submitted to *Planning Science Education at the Secondary Level.* Johannesburg: IIEP and CEPD

Ndunda, M & Munby, H. 1991. Because I am a woman: A study of culture, school, and futures in science. *Science Education*, 75(6), pp 683–99

Odaga, A. & Henneveld, W. 1995. *Girls and Schools in sub-Saharan Africa: From Analysis to Action.* World Bank Technical Paper. Washington DC: World Bank

Ogunniyi, MB. 1995. The development of science education in Botswana. *Science Education,* 79(1), pp 95–109

Ogunniyi, MB. 1986. Two decades of science education in Africa. *Science Education*, 70(2), pp 111–22

Okebukola, P. 1995. Organisation and conditions of secondary science education in Nigeria. Submitted to *Planning Science Education at the Secondary Level.* Johannesburg: IIEP and CEPD

UNECA (United Nations Economic Commission for Africa). 1990. *The Situation Analysis and Strategies for the Promotion of Girls/Women to Scientific and Technical Training and Professions.* Technical publication. Paris: UNESCO

Wasanga, PM. 1995. Science education in Kenya. Submitted to *Planning Science Education at the Secondary Level.* Johannesburg: IIEP and CEPD

*West Africa Weekly Magazine*, 26 June to 2 July, 1989, p 140

Zonneveld M, Taole J, Nkhwalume A & Letsic, L. 1993. The mathematics classroom: Interaction and distraction. First annual SAARMSE Conference proceedings. Rhodes Unversity: Grahamstown

Zimbabwe. Ministry of Education. 1995. The status of science education in Zimbabwe. Submitted to *Planning Science Education at the Secondary Level.* Johannesburg: IIEP and CEPD

# 7

## Teacher education: Pre-service and in-service support models

*Hubert Dyasi, City College, City University of New York, and Karen Worth, Education Development Center and Wheelock College, Boston, Ma, USA*

### ABSTRACT

The goals of science and technology education demand the implementation of good teacher development programmes. This chapter examines teacher education and support models for pre-service and in-service education used in the past and present. The authors analyse the curriculum for science teacher education; support structures such as materials, finance, and teachers' centres; relationships between schools and teacher education institutions; and teacher educators and their professional development. Importantly, this chapter delineates alternative paradigms for teacher development for the future.

### INTRODUCTION

At a teacher training college in Zanzibar, Tanzania, selected secondary school teachers, principals, three scientists, five teacher educators, and 60 secondary school students participate in a three-week residential camp focusing on professional development in science. In New York City, USA, for two weeks during the summer vacation, 45 teachers and assistant principals attend a professional development programme, and meet for three hours on Thursday evenings throughout the following school year of two 11-week terms. On five consecutive Saturdays, beginning in the fourth week of the first term, they are joined by 75 primary school students who have volunteered to learn science through inquiry under the direction of teachers in the programme. Just outside Durban, South Africa, teachers participate in a constructivist-based two-year professional development programme that integrates science, science teaching and learning; resource and management skills; and delivery of professional education to other teachers. Working as members of a professional team of teachers and staff from universities and industry, full-time science education students in Ghana are assigned to an industry where they learn science by practising it. Subsequently, they and their professional team prepare and produce science education resource materials for use by teachers. At the University of the Western Cape, South Africa, undergraduates work with teachers to organize and conduct an annual 'young scientists' competition' for pupils from local schools. Similarly, architecture undergraduates at the City College, New York and teachers in nearby schools enhance their professional growth by working together one day a week to engage students in constructivist-based learning about the 'built environment'.

We know these programmes through designing and implementing our own and other professional development programmes, interacting with colleagues, and from the educational literature (Keohane, 1974; Ramsey, 1974; Van der Cingel & Yoong, 1979; Harlen, 1979; and Power, 1988). Though different, they share common themes, such as general principles of practice, and mechanisms that enable each programme to fit its local contexts. These provide a unity in the diversity of the programmes.

In this chapter we discuss illustrations of this idea of unity in diversity to highlight the fact that, although the quality of classroom practice is related to a teacher's professional knowledge base and skills, contextual factors intervene to present opportunities for and obstacles to its utilization (Darling-Hammond & Goodwin, 1993; Carnegie Task Force, 1986; Harlen 1993). Professional development is critical for the development of that knowledge base and skills, but its structure and design are deeply dependent on contextual factors.

We first refer to a general knowledge base for teachers and raise contextual questions relevant to the development of science teachers. We then examine a variety of professional development mechanisms and strategies. A discussion of resources highlights their importance in determining the quality and sustainability of science teacher education programmes. Finally, we refer briefly to programme assessment only to suggest that teacher development programmes should be assessed. The design, implementation and assessment of teacher development programmes merit concerted inquiry.

We ask readers to keep in mind the unity-in-diversity notion and to think of ways to adapt and use the illustrations we present to suit their own contexts. We do not suggest a royal road to excellent professional development programming. Rather, we present examples of how some science educators have examined their own situations and used general principles to design and implement suitable programmes. Only a judicious alignment of contextual factors with pervasive principles of professional development leads to the successful planning and implementation of effective programmes for science teacher development.

## ELEMENTS IN THE DESIGN OF PROFESSIONAL DEVELOPMENT PROGRAMMES

### A knowledge of teaching and learning science

There is a broad knowledge base that forms the foundation for science teaching. Shulman (1987: 8) identified the following components of this knowledge base:

 

content knowledge; general pedagogical knowledge such as teaching strategies, classroom management and organization; knowledge of educational ends, purposes, values, and their philosophical and historical bases; pedagogical content knowledge — a special blend of content and pedagogy; a knowledge of curriculum materials and programmes; a knowledge of learners and their characteristics; and a knowledge of educational contexts, such as school governance and the characteristics of communities. We would include a knowledge of assessment of classroom instructional approaches, learning experiences, and of student progress.

## The local context

In using the knowledge base components to design professional development programmes for teachers of science, one must consider the local education policy and how it can be reflected in the programme. An education policy that requires all students to study science at all education levels demands different programmes from those designed for a policy that prefers selection of the highest-achieving students.

One must also consider the education system's vision of excellent science education. A vision might be that students acquire science concepts through lectures and laboratory activities designed to yield only one correct answer, or of an acquisition of science concepts and practice through engagement in science inquiry. The vision might be the dictation of science facts to enable students to pass examinations for admission to tertiary institutions or for acquisition of a general knowledge of many science topics. A curriculum developer may think a vision is inadequate and needs to be changed. If so, the developer must determine how the system's vision for the professional development of teachers of science relates to the major professional components deemed necessary. In many instances the vision is not explicitly stated but can be inferred from professional development curricula, syllabi and examinations. In some cases the vision is explicitly stated as standards for the professional development of teachers. Given a vision of the desired science education, the designer of professional development programmes must consider: (1) suitable professional development mechanisms; (2) the demands of the vision of science education and of professional development; (3) lessons learned from strategies used by other programmes; (4) the degree of discontinuity between ingrained local practices and habits of mind and the desired successful mechanisms — it may be necessary to choose a less-than-ideal mechanism to reduce such discontinuities; (5) the availability of the requisite human and material resources for the mastery and implementation of the mechanism; (6) the characteristics of participants for whom the programme is intended; and (7) structural constraints such as scheduling requirements.

One must consider the strategies that can be used to realize the vision. While mechanisms are static features within which professional development activities are carried out, strategies are dynamic processes for achieving desired ends. For example, within the fixed structure of a workshop, developers might adopt a 'hands-on strategy' or a 'story-telling strategy', depending on their mindset or the constraints of the local situation.

 

One must consider the resources that are available for the desired professional development programme. In addition to considerations of the education of the teachers, one considers human resources in the community, such as exemplary science teachers who can serve as mentors, appropriately qualified science educators, scientists, and other learning specialists. One must also consider physical facilities, equipment, materials, supplies and schemes available for assessing the effectiveness of the programme.

All these issues are important. We have, however, chosen to focus on structure strategies, resources, and assessment. Questions of policy, and visions of science education and of the professional development of teachers of science are best addressed in the context of specific countries.

## PROFESSIONAL DEVELOPMENT STRUCTURES

We arbitrarily divide professional development structures into two complementary categories — formal and informal.

### Formal structures

Formal structures have a set schedule and development of teachers' science education knowledge as a goal.

They can be can be two- to three-week workshops, institutes or camps offered during school vacations, such as the Zanzibar Science Camp. For four hours every morning the participating adults work in teams with students to teach science through inquiry. At the conclusion of the morning lessons, each team reviews, discusses, analyses and assesses the instructional approach as well as their own professional development during the camp sessions. During the afternoon all participants are exposed to development workshops that include demonstrations by staff, while students work with computers and receive English instruction. In the evenings, participants prepare lessons for the following days. Camp resource staff consists of a ministry of education educator who serves as the camp administrator, scientists, science educators, teachers, training college tutors and school inspectors. Throughout the following school year, camp staff visit participant teachers in their schools as they implement the approach they learned at the camp. In each zone, teachers meet in clusters coordinated by colleagues selected for their enthusiasm and released from teaching duties on a part-time basis.

Few places have academic structures that use a two- to three-week institute during school vacations, followed by weekly or monthly sessions during weekends or after school. When used, the model usually requires participating teachers to complete a two- to three-year sequence of sessions for a professional qualification.

The Council for the Advancement of Science and Mathematics Education (CASME) in Durban, South Africa, uses such a mechanism. During vacations, CASME staff conduct high-school teacher-leader workshops that integrate science understanding with a constructivist-based science teaching approach. CASME sustains participants' education during school terms through distance learning. Teachers must participate in the programme for two years for successful completion. In Harlem, New York, at the City College of the City University of New York, about 45 teachers and assistant principals take part in a two-week summer institute designed to deepen their knowledge of science and of science teaching using an inquiry approach. During the following academic year, they attend a professional development programme, meeting three hours a week after school for 24 weeks. On five consecutive Saturdays, beginning in the fourth week, they are joined by 75 primary school students who have volunteered to learn science through inquiry under the direction of the programme participants. During these Saturday sessions, teachers divide into teams of five to teach science through inquiry to ten students per team. Each student is accompanied by a parent who participates in a three-hour parents' workshop conducted by participating teachers on inquiry in science and on the parent's role in helping a child learn science at home. After each Saturday session, teachers and staff spend an hour discussing and analysing the morning and plan for the following Saturday. City College staff, consisting of a science educator who serves as leader, two teachers, a scientist, and a parent-education specialist, spend an additional hour reviewing the day's events. During the second half of the school year, teachers use the teaching approach and science they have learned in their classrooms. During the three-hour weekly sessions, they function as a study group discussing their teaching and experiences.

The placement of teachers in university or industrial science laboratories is not a widespread professional development mechanism.

At the University of Cape Coast in Ghana, full-time education students work as members of a science materials development team of university and industry staff and teachers. Their participation involves working in an industry, learning science through its practice, and using the science knowledge thus gained to help their professional teams to prepare science education resource materials. To enhance secondary school science teachers' understanding and teaching of science, the Columbia University College of Physicians and Surgeons, New York, conducts a two-summer vacation programme in which teachers serve as members of laboratory research teams led by university staff. Teachers see new avenues for their personal and professional growth, revitalize their science teaching, increase laboratory-based participatory learning in their classrooms as well as their capacities to communicate the excitement of science to their students and fellow teachers. Through its Science Outreach Program in New York, that encompasses a summer vacation course and year-long academic activities, the Rockefeller University provides laboratory-based research experiences for high school students and their teachers. The programme teaches students the culture and ethos of scientific investigation, science content and process skills. In partnership with scientists, teachers serve as mentors to students, thereby enhancing their own skills at reviewing students' notebooks, and asking questions to promote their laboratory work, scientific writing, and oral and aural skills.

Another professional development structure is three- or four-year programmes designed for teachers or prospective teachers. Such a programme usually requires enrolment in a range of courses in arts, sciences and education.

In a programme for primary school teachers at Wheelock College in the United States, the arts and science courses are taken concurrently with courses in human development, children's learning, curriculum design, developmental pedagogy and the observation and assessment of learning. Together, these courses lay a foundation for participants' knowledge of science pedagogy and their subsequent clinical experiences. The science courses are developed and co-taught by a life scientist and a physical scientist, and provide a two-semester foundation course that is taken by students during their first and second years. The course provides them with a foundation for, an understanding of and positive attitude towards science. During the third or fourth year, most students enrol in a course called 'Teaching science to children' that provides students with a basic understanding of science pedagogy. Students entering this course already have an understanding of basic science concepts and how they arise as a result of science inquiry. They also have an understanding of child growth and development as well as of how children construct meaning through direct experience, social interaction, and their own reflection and thought.

Clinical experience is an essential component of an extended professional development programme. To create a suitable environment an institution such as Wheelock College builds a close relationship with local schools that serve as professional development sites. These sites enable Wheelock students to engage in practice teaching for two 14-week periods. A weekly teaching seminar accompanies the field work, frequently co-led by a teacher and a professor. During the seminar students reflect on their experiences in classrooms, continue to learn to inquire into their own practice, and are supported in observing and assessing the learning of children.

University degree programmes designed for secondary school teachers may require that students major in a science subject as well as in education, as is the case at the University of Cape Coast in Ghana or at Njala College at the University of Sierra Leone. In other cases, teachers major in a science subject and then complete a one-year higher education diploma.

Finally, some education systems conduct one-off workshops or lectures to keep teachers abreast of teaching methods, introduce them to a new technology or to meet contractual requirements. Although organized as professional development, in most cases these activities serve dissemination purposes.

### Informal structures

There is a large variety of informal structures for the professional development of teachers of science. These are more ad hoc and often planned and led by teachers.

A growing informal structure is the provision of mentor–teachers to groups of schools. In one structure, on request, mentor–teachers spend the entire day at a school. The mentor–teacher works with four or five teachers who have been released from their classrooms. They first discuss key aspects of a science lesson the mentor–teacher will conduct. During the lesson, the class teacher co-teaches while the others observe and take careful notes. After the lesson the mentor serves as facilitator as teachers discuss their observations and interpretations, and raise questions regarding the implementation of the approach in their own classes. During the discussion, the mentor highlights points that epitomize the teaching approaches used and how they were evident in the lesson. A City College Workshop Center science educator demonstrated the practicability of this structure in Cape Town, where a few teachers combined their classes so that they could all observe her teach and discuss the lesson.

Most African countries conduct high school leaving examinations in science subjects. The West African Examinations Council, for example, conducts such examinations in Ghana, Nigeria, The Gambia and Sierra Leone. Science teachers from each country administer the practicals and form teams to mark and score examination scripts. This involvement becomes an informal professional development mechanism. As markers, science teachers enhance their knowledge of the science topics covered in the examinations and benefit from interactions with colleagues from other schools and countries.

Science teachers engage in professional development when they attend conferences, where they enhance their knowledge of science, science teaching and other professional aspects of science education. Similarly, teachers develop professional skills when they participate in the development of science curricula or learning materials. The African Primary Science Programme (APSP) and the Science Education Programme for Africa (SEPA) used this mechanism in long summer workshops attended by participants from several continents. During the following school year, participants tested materials produced at the summer workshops and adapted them to their own circumstances. Science teachers in Nigeria benefited professionally by participating in the curriculum development work of the Science Teachers' Association of Nigeria (STAN). The Caltech Pre-service Science Initiative in California has teams of teacher–leaders who develop science education materials for use in the professional development of science teachers through a sequence of teacher development workshops. All these teacher development mechanisms involved scientists as members of the resource teams.

Scientists and science educators can become involved in other ways in the professional development of teachers. Teachers' knowledge is enhanced when scientists visit schools to demonstrate and lecture to students. Teachers grow when they work with scientists and university undergraduates to help high-school students develop projects for science fairs, or when they judge exhibits in other schools. University students in the 'built environment' programme, discussed earlier, provide technical assistance to teachers as they teach their students the technical skills and knowledge necessary for a successful *charrette* — a parallel activity to the science fair.

## Professional development strategies

Teacher-educators can choose from an array of strategies. Their choices will be influenced by their beliefs about the nature of science and how human beings learn, as well as by local factors such as the availability of resources and the number of teachers. We preface a discussion of some key strategies by stating our own beliefs.

### *Our beliefs*

Adults as well as children learn through direct experience, inquiry and reflection, and interaction with their cultural traditions (see Hawkins, 1976; Driver, 1985; Resnick, 1987; Yager, 1991; Dyasi, 1992). Science is learned by engaging in learning activities that bear fidelity to the nature of science and to the ways that science generates knowledge. Through the development of their own knowledge base, learners come to know first-hand how scientific facts, concepts, laws and generalizations are acquired and established. By engaging in science as inquiry teachers directly learn the value of: (1) open-ended and continuing investigations and studies; (2) collaborative learning groups; (3) a research group revisiting an investigation; (4) reporting to critical but friendly inquirers who know the value of impersonal criticism; (5) exploring an idea for a 'research conference'; and (6) generative discussions in the 'research conference'. By engaging in these activities, teachers strengthen their knowledge of science content, of laboratory techniques and equipment, of participation in scientific discourse, and of how to use relevant resources.

The principles that underlie the professional development of science teachers apply equally to teacher preparation and in-service education. Teachers acquire the requisite knowledge, beliefs, and skills during their teacher-preparation phase, and continue to deepen their knowledge when they become classroom teachers.

In some situations, a university or a teacher training college might be best suited for prospective and in-service teachers; in other situations the most suitable provider may be a school, a department in a ministry of education, a teachers' collaborative or a combination of institutions. However, in all cases, the nature of the content and learning is critical — both must exemplify the theoretical and practical knowledge and skills needed for effective science education in classrooms.

## Characteristics of effective strategies

### *Integrated staffing*

The selection, composition, development and functioning of staff is probably the most important strategy in any professional development effort. Professional development staff must have a clearly defined and shared vision, the necessary qualifications, an exemplary history of practice of their own speciality, and the courage, energy and support by the system to implement their vision. Professional development staff should work as an integrated unit that ideally should consist of professionals such as teachers, scientists, science educators, and education supervisors who complement one another.

### *Learning through models*

In the best professional development programmes, staff model what teachers should do in classrooms. If teachers are to teach through science inquiry, they should be taught through scientific inquiry. Duckworth's (1986) work on developing inquiring teachers has provided interesting data. Dyasi (1992) has also described science professional development programmes that engage teachers as inquirers.

### *Customized programming*

Teachers and prospective teachers have different needs that cannot be met by a one-size-fits-all approach. The provision of professional development must be differentiated according to the groups served. Teachers of young children, for example, need exposure to an approach that emphasizes the methods scientists use to uncover a phenomenon and the ability to conduct open-ended inquiry, rather than memorization of theoretical constructs. High-school teachers' education, however, may, in addition, focus on learning how to engage students to use inquiry in order to internalize scientific concepts and unifying principles. Professional development programmes must take account of such variations. For example, a teacher might be skilled in designing laboratory experiences but be inadequate at conducting productive student discussions of those experiences.

### *A focus on science*

If teachers are to guide students towards an understanding of the nature of science they must first understand it themselves. There is general agreement on what science is, but different places and even schools emphasize different aspects. In the United States, for example, the National Science Education Standards highlight: (1) unifying concepts and processes that cut across categories of content; for example, systems, order and organization; evidence, models, and explanation; change, constancy and measurement; evolution and equilibrium; and form and function; (2) science as inquiry, and as a combination of processes and knowledge required to understand scientific reasoning and knowledge; (3) physical science, life science, and earth and space science comprise the subject matter of science; (4) science, technology and decision making as a connection between the natural and designed worlds; (5) science in personal and social perspectives, and the use of science to understand and act on personal and social issues; and (6) the history and nature of science in a way that reflects its development and ongoing nature. All science teachers, therefore, should not only have an understanding of these components, but should come to that understanding through first-hand participation in practical learning in a manner that is consistent with research-based principles. But many science curricula emphasize the subject matter almost to the exclusion of the other components. For our part, we tend to emphasize science as inquiry because, in the words of the US Standards,

 

> [i]nquiry is a multifaceted activity that involves making observations; posing questions; examining books and other sources of information to see what is already known; planning investigations; reviewing what is already known in the light of experimental evidence; using tools to gather, analyze, and interpret data; proposing answers, explanations, and predictions; and communicating the results. (National Research Council, 1994: 23)

Thus in schools, science inquiry refers to the activities of students in which they develop a knowledge and understanding of scientific ideas as well as an understanding of how scientists study the natural world.

### *Systemic support*

Systemic support implies concerted contributions by different groups to the provision of high-quality science education at all education levels through professional development programmes for science teachers. Systemic support requires the mobilization of groups such as teachers, educational administrators, parents, scientists, policy makers, examining bodies, nongovernmental development agencies, education development organizations and foundations, financial institutions, and business and industry. Without their combined support, the effective planning, design, financing and adoption of professional education programmes remains elusive. The differing interests, energies and resources must be orchestrated to create mutually supportive relationships that sustain efforts to establish and maintain quality science education in schools. Systemic support is especially important when introducing new programmes, as it can allay fears and uncertainty about what works and what does not as systems change (see Fullan & Miles, 1992; St John, 1991).

### *Clinical and regular classroom experiences*

Observing and understanding students' behaviour is an important component of teaching. Participants in professional development programmes should spend time in schools studying students learning science individually, in small groups, and in regular class sizes — in their own and others' classrooms — so that their learning of teaching strategies, and of when and how to implement them, is founded in the worlds of children and schools. In those worlds, they should watch expert teachers support and guide learning and how children make sense of their world. If they are entering teaching, they may try out their own skills, initially with small groups of children, then later with whole classes.

### *Inquiry into practice*

A good teacher must continue to gain knowledge beyond that acquired in basic teacher preparation programmes. Teachers should keep their practice up-to-date and consistent with the development of the profession. A useful strategy for achieving this in development programmes is to encourage teachers continually to inquire into, and make meaning of, their own classroom practice. Thus, habits of inquiry become ingrained, such as the collection of data to assess teaching and learning to improve decision making about the selection of learning activities, teaching procedures and student learning. Such programmes forge strong links between professional practice and current professional knowledge. Don Schön's work on the reflective practitioner demonstrates the importance of this component of professional development. Other scholars whose work gives prominence to teacher inquiry are Kemmis and McTaggart (1981), who developed a model to involve teachers in studying and documenting their practice in cycles of planning, acting, observing, reflecting, re-planning and so on. Lieberman (1995) and Lieberman and Miller (1992) have contributed immensely to teacher participation action research.

As they choose strategies, those who reform teacher education programmes must grapple with hard questions. Some examples follow.

*When teacher preparation takes place in a university, should there be separate science courses for students who will become teachers? Should the content of science courses for teachers be explicitly linked to what children should learn in science, and how?*

We assume that in either case science courses are taught by professionals who teach through inquiry and convey an image of science that is inquiry-based, and that students have an opportunity to engage in inquiry.

*Should science courses for teachers be broad overviews that provide a glimpse of the field and the nature of the field? If so, what content should be covered? Should all teachers have in-depth experience in a science or only teachers of older children?*

Teachers, especially at the primary level, are expected to teach across a number of domains. Therefore, some argue that teachers should have a broad view of science. Others emphasize experiencing a science in depth, arguing that only by doing so can teachers understand the nature of scientific investigation, and that further knowledge can be acquired later as needed.

*Should knowledge of how to teach science be delivered exclusively through workshops and courses on methods? Does the methodology emerge from reflecting on and learning from teachers' own science learning?*

Some argue it is sufficient if science courses model what should go on in classrooms. They argue that any discussion of teaching and learning that links the coursework to classrooms distracts teachers from the science. Others suggest that learning to reflect on one's own learning as a science learner is critical. Still others suggest that engaging teachers in a continuing dialogue between learning science and teaching students is a more effective way of building knowledge of science teaching.

*How should college or university science professionals collaborate with science education specialists and teachers? How should science courses relate to those in science teaching?*

Most argue that communication across fields is valuable and important. Clearly the science teaching modelled in science courses should be based on the same beliefs about science and learning that teachers are exposed to in their science education courses. It is less clear whether there should be a connection between the content of the two domains.

*What is the appropriate balance in professional development at different times in a teacher's career between time spent learning about how to teach science in a workshop or institutional course structure and learning through working with pupils and teachers in classrooms?*

More traditional programmes emphasize coursework and skills training to a greater extent than field experience. However, today many argue that the clinical component of pre-service education must have equal attention, since classrooms are where prospective teachers learn how to use their knowledge. Increasing its prominence has implications for supervisors and cooperating teachers who must become facilitators of this learning. The divergence is less about importance than about how best to use time, a scarce resource.

*How much emphasis should be placed on learning to inquire about pupils and teaching?*

Many argue that this is a new and important component of science education reform that should be part of all teacher education programmes. Teaching for understanding, whether of science or other domains, requires understanding pupils' learning and the relationship between teaching and learning. The ability to reflect and inquire about practice and pupils' learning must begin at the pre-service level.

## RESOURCES FOR THE PROFESSIONAL DEVELOPMENT OF TEACHERS

### Human resources

A wide range of human resources is required for the design and implementation of professional development programmes. These resources encompass teachers, research and teaching scientists, science educators, school administrators, teaching and learning specialists, curriculum developers, students, and education staff of funding institutions. No programme uses the complete range; instead, various combinations of the different human resources are used.

Teachers function as professional educators when they serve in programmes designed for the continuing education of other teachers and when they supervise clinical experiences of prospective teachers in their classrooms. During the development of the African Primary Science Programme (APSP), highly talented teachers were involved in the development of curriculum materials, thereby advancing their own classroom practice. They later joined the staff of professional development programmes. In the United States, a professional development programme for high-school science teachers, funded by the Woodrow Wilson Foundation, first identified outstanding science teachers and employed them as leading staff members in vacation workshops for high-school science teacher leaders. The workshops are held on university campuses in different parts of the country. More than 15 years ago a group of primary-school teachers in Philadelphia formed a teachers' learning cooperative to provide professional development for one another and for other teachers interested in facilitating pupils' learning. They selected topics of interest and took turns in leading professional development sessions that often consisted of sharing ongoing classroom work and discussing its meaning and research base. Sometimes they invited a specialist as an observer, commentator, or occasionally as the presenter of a selected topic. In a different situation, a high-school science teacher teamed up with a research scientist at the National Observatory in Cape Town, South Africa, to conduct physics teaching demonstrations for high-school teachers.

Teachers can provide concrete, authentic, personal experiences of how students benefit from the learning approaches advocated in professional development programmes. They can portray a realistic rather than an ideal picture of how the approaches can be adapted to suit classroom situations that involve ordinary students in ordinary schools. And, because they are respected colleagues, teachers have more credibility than science educators and scientists who are distant from the realities of schools.

Staff of practically all professional development programmes for science teachers include a science education specialist. Indeed, in most cases a science educator serves as designer and leader of the programme. All science educators have specialized knowledge and experience of educating teachers and other professionals in the school system. Apart from providing a perspective from prior experience as a science major at university and as a school science teacher, a specialist science educator brings research-based knowledge of learning and curriculum development to the programme. Such a person can also provide the knowledge and practice of assessment.

In addition to their knowledge of their science specialties, scientists continually practise scientific inquiry and research as part of their development as professional scientists. Because one of the goals of a science teacher development programme is to enable participants to acquire science knowledge and concepts through science inquiry, a scientist is an essential human resource.

In some programmes mentioned in this chapter, university students assist teachers to carry out special learning activities in schools — such as investigations for science fairs. The university students are often more up to date than teachers in their knowledge of science subject matter and of research-based science learning practices, but do not yet have the necessary experience to apply their knowledge in the classroom. Thus, a professional education partnership can develop that benefits both the college student and the teacher.

Perhaps the most deficient aspect of the professional development of science teachers is the paucity of programmes for science teacher educators. Professional development programmes include experienced teachers, curriculum developers, science subject specialists, inspectors, and specialists in science-related fields. Each group carries excess baggage from its field of study. For example, even though they have experience of teaching high-school classes, science teachers' visions often match those of traditional college teaching. School inspectors who have been principals are better at administration than at science teaching. Experienced primary-school teachers may have excellent knowledge of working with pupils but lack a knowledge of science. Younger teachers with a bachelor's degree in a science may not have enough classroom teaching experience. Scientists may provide technical science information but lack a knowledge of how children learn. Such groups may be enthusiastic and concerned professionals, but they need exposure to specialized development programmes in the professional education of teachers of science.

There are programmes for the further education of such teacher–educators. With support from the Commonwealth Secretariat, the United Nations Economic Commission for Africa (UNECA), and the United States Agency for International Development (USAID), the Science Education Programme for Africa (SEPA) established and implemented a programme that created a core of science educators in participating countries. Scientists need orientation if their input to science teacher professional development programmes is to be maximized. They need to participate in scientist orientation activities, as is done by the California Institute for Technology in collaboration with the Pasadena School District, and the Merck Institute for Science Education, USA, in association with four school districts. The New York Academy organizes similar orientation programmes for scientists together with New York city school districts. Scientists in many African institutions of higher learning work with teachers on school science improvement.

## Material resources

Effective professional development programmes require an appropriate physical setting with adequate material resources. The setting must be recognized by teachers and the education system as providing the necessary continuity. It must be reasonably well equipped and could be a school, a teachers' centre, or a curriculum development centre. The setting must promote the creative use of resources and interactions between teachers and staff. As the setting continues to accumulate resources and participants' work, its users will assume ownership, making it an accessible intellectual and professional resource.

The setting should have curriculum resources, science teaching kits, a library of exemplary teachers' work and lessons, students' notebooks, journals, and examples of assessment instruments. Countries such Ghana, Kenya, Malawi, Nigeria, Sierra Leone, Tanzania and Uganda had a tradition of well-supported curriculum development centres which served as nodes for professional development programmes. That tradition included a level of respect for the professionalism of teachers and support, such as networks of teachers' centres.

Completion of a long-term professional development programme should lead to recognized certification or salary increments. Teachers attending short after-school, weekend or vacation workshops should be given financial support for travel, boarding and incidental expenses. On completion of a professional development programme, teachers should be given the necessary support materials for classroom implementation.

Every programme needs funds or contributions in kind for staff remuneration, and the nature of the remuneration should depend on local practice. For example, a university or industry might give its staff paid release time to work on teacher development programmes or it might award tuition vouchers to teachers who serve as supervisors of clinical experiences. The institution conducting the development programme may procure funds from industry, foundations or the local education authority to employ scientists, science educators and teachers as adjunct staff.

National, provincial and local government education budgets must include support for the professional development of science teachers. But responsibility for support for such programmes cannot be left to the government alone. Business and industry, institutions of higher learning, and philanthropic foundations should support systemic change programmes that sustain development.

Resources for national science education development programmes vary, depending on the relationship between the supply of teachers and the demand for them. When the demand is high and the supply short, enrolment in development programmes is high, facilities become strained and courses are shortened, as happened in much of Africa immediately after independence. Professional education of teachers of science differs when countries have different resource bases. In such situations, regional science education development programmes can play a significant role in maintaining vision. Countries in Africa can benefit by sharing knowledge, experience and human resources on programme development, and by collaborating on a regional basis. Lessons learned in countries outside Africa can be adapted by regional organizations working in partnership with professionals from different African countries. In this context, the work of the African Forum for Children's Literacy in Science and Technology (AFCLIST) is invaluable.

## ASSESSMENT OF PROFESSIONAL DEVELOPMENT PROGRAMMES

Professional development programmes must be assessed to ensure that the views of science and of learning they portray are consistent with programme goals and objectives. Their learning activities must be assessed to ensure their effective contribution to participants' learning and self-confidence about activities such as the design and implementation of learning, their knowledge of science content, and how to conduct investigative science lessons. The range of strategies used by professional development programmes must match the characteristics of the teachers they serve as well as other contextual factors.

Teacher growth along the dimensions identified provides indicators for programme assessment. For example, if an objective is the development of teachers' capacities to design, carry out, and make sense of investigations of natural phenomena, assessment would generate appropriate data by continual engagement of teachers in that activity. If the programme aimed to develop teachers' abilities to conduct investigative science lessons, then assessment would include analysis of teachers' decisions on the choice of learning experiences, classroom observations, teacher interviews regarding teachers' roles, students' roles, and the role of selected instructional materials, collection of student portfolios, and determination of the adequacy of provisioning for investigative science learning. Similarly, assessment of science teacher educator programmes would address participants' growth as excellent educators.

## TOWARDS THE FUTURE

We began by proposing that only a judicious alignment of contextual factors with pervasive principles of professional development can lead to effective programming. Achieving such an alignment requires careful analysis. If the analysis is inadequate, or too strongly based on factors other than on professional realities, teachers may become frustrated, education can suffer and society can experience disillusionment. We claim that with systemic support — and by this we mean a unity of vision of curriculum developers, teacher educators, examination staff and so on as much as provision of resources — pervasive principles of teacher development can be applied. We contend that doing so enables teachers to become reflective practitioners able to do the best possible within their contextual constraints, be they those of well-equipped private schools in industrialized countries or the large, poorly resourced classes found in much of Africa.

## REFERENCES

Carnegie Task Force on Teaching as a Profession. 1986. *A Nation Prepared: Teachers for the 21st Century.* Washington, DC: Carnegie Forum on Education and the Economy

Darling-Hammond, L & Goodwin, A. 1993. Progress toward professionalism in teaching. In Gordon Cawelti (ed). *Challenges and Achievements of American Education: 1993 Yearbook of the Association for Supervision and Curriculum Development,* pp 19–52. Washington, DC: Association for Supervision and Curriculum Development

Duckworth, E. 1986. Teaching as learning. *Harvard Educational Review,* 56(4), pp 481–95

Dyasi, HM. 1992. Developing confidence in primary-school teachers to teach science and technology — a practical approach. In David Layton (ed). *Innovations in Science and Technology Education,* vol IV, pp 23–38. Paris: UNESCO

Fullan, MG & Miles, MB. 1992. Getting reform right: what works and what doesn't. *Phi Delta Kappan,* 73(10), pp 744–52

Harlen, W (ed). 1993. *Education for Teaching Science and Mathematics in the Primary School.* Paris: UNESCO

Harlen, W. 1979. Towards the implementation of science at the primary level. In Reay, J (ed). *New Trends in Integrated Science Teaching,* vol V, pp 59–67. Paris: UNESCO

Kemmis, S & McTaggart, R. 1981. *The Action Research Planner.* Victoria, Australia: Deakin University Press

Keohane, KW. 1974. The preservice education of teachers of integrated science at training colleges and universities. In Richmond, PE (ed). *Trends in the Teaching of Integrated Science,* vol III, pp 53–9. Paris: UNESCO

Lieberman, A. 1995. Practices that support teacher development: Transforming conceptions of professional learning. *Phi Delta Kappan,* 76(8), pp 591–96

Lieberman, A & Miller, L. 1994. The professional development of teachers. In Atkin, M, (ed). *The Encyclopedia of Educational Research,* 6th ed, vol 3, pp 1045–53. New York: Macmillan

National Research Council. 1994. *National Science Education Standards.* Washington, DC: National Academy Press

Power, C. 1988. New methods for training and retraining science and technology teachers. In Layton, D (ed). *Innovations in Science and Technology Education,* vol II, pp 283–95. Paris: UNESCO

Ramsey, G. 1974. The in-service education of teachers of integrated science. In Richmond, PE (ed). *New Trends in the Teaching of Integrated Science,* vol III, pp 60–8. Paris: UNESCO

Resnick, L. 1987. *Education and Learning to Think.* Washington, DC: National Academy Press

Schön, D. 1987. Educating the reflective practioner: towards a new design for teaching and learning in professions. San Francisco: Jossey Bass

Shulman, LS. 1987. Knowledge and teaching; foundations of the new reform. *Harvard Educational Review,* 7(1), pp 1–22

St John, M. 1991. *Science Education for the 1990s: Strategies for Change — Reflections on a 1991 Wingspread Conference.* Inverness, Ca: Inverness Research Associates

Van der Cingel, N & Yoong, CS. 1979. Education of teachers for integrated science. In Reay, J (ed). *New Trends in Integrated Science Teaching,* vol V, pp 87–99. Paris: UNESCO

Yager, Robert E. 1991. The constructivist learning model: towards real reform in science education. *The Science Teacher,* 58(6), pp 52–7

 

# 8

## Teaching large classes

*Gilbert Onwu, University of Ibadan, Nigeria*

### ABSTRACT

After the adoption of the principle of universal primary education, the 1970s and 1980s saw an unprecedented expansion of student enrolment in African countries. As a consequence, class sizes have increased dramatically, with a concomitant decrease in the quality and quantity of resources. This chapter discusses teaching large classes in a context of poor resourcing. It examines the reality of large classes; policy and practice issues; the impact on the quality of learning in large classes; what research is available on teaching large classes; resource utilization; and innovative approaches in teaching large classes.

### INTRODUCTION

An analysis of education in low- and middle-income countries of Africa reveals compelling problems as well as substantial accomplishments. At independence, in a determined bid to make formal education more accessible, many African countries embarked on far-reaching educational programmes premised on the philosophy of 'education for all'. In these countries, Universal Primary Education (UPE) became a major policy thrust. An inevitable feature was an unprecedented expansion of educational systems over one or two decades. Both pupil enrolment figures and pupil–teacher ratios increased dramatically.

World Bank figures based on a study of education in sub-Saharan Africa (World Bank, 1988) show that between 1960 and 1983 the number of primary pupils expanded by about a factor of four and the number of secondary pupils by a factor of 14. For example, in Nigeria, the country with the highest enrolment rates in Africa (Bajah, 1995), the primary school population rose from 6,6 million in 1976 to 13,6 million in 1990, while secondary school enrolment shot up from 0,6 million in 1976 to 3,8 million in 1990. This growth rate can be attributed to the implementation of UPE in 1976 and the launch of the National Policy on Education (NPE) in 1982.

Recent statistics from Nigeria's Federal Ministry of Education show that in 1994 there were 360 782 teachers and 18 296 202 pupils (a teacher–pupil ratio of about 1 : 50) in 39 221 primary schools. Figures are not yet available for secondary education, but the estimated enrolment figure for 1995 is about 3,5 million pupils. At the tertiary level, total student enrolment in the universities stood at 77 481 in 1981 but soared to 224 879 in 1992.

The demand for formal education, with a concomitant increase in school enrolments, has resulted in a dramatic increase in class sizes, with attendant high teacher–pupil ratios. The Nigerian situation reflects that in most African countries.

## CLASS SIZE AND FUNDING

A recent United Nations Children's Fund (UNICEF) report, *The Progress of Nations* (1994), highlights strikingly divergent class sizes in the world's primary schools, varying from about 12 in Norway and Sweden to over 90 in the Central African Republic. Although pupil–teacher ratios have remained relatively stable over the last decade, a cursory look at table 8.1 shows that class sizes increased in some African countries in the 1980s. Generally, class sizes in developing countries are two to four times larger than in industrialized nations.

For many African countries high enrolment is an increasingly serious problem (Ajeyalemi et al, 1990). As enrolments have increased, annual government spending per pupil at all levels of education has fallen (see table 8.2).

**Table 8.1: Number of pupils per teacher by country, 1980 and 1990**

| Country | 1980 | 1990 | Increase (%) |
|---|---|---|---|
| Burundi | 39 | 67 | 28 (72 %) |
| Central African Republic | 60 | 90 | 30 (50 %) |
| Senegal | 46 | 58 | 12 (26 %) |
| Bolivia | 20 | 25 | 5 (25 %) |
| Oman | 23 | 28 | 5 (21 %) |
| Bangladesh | 54 | 63 | 9 (17 %) |
| Pakistan | 37 | 43 | 6 (16 %) |
| Lesotho | 48 | 55 | 7 (15 %) |
| Mauritius | 41 | 47 | 6 (15 %) |
| Congo | 58 | 66 | 8 (13 %) |

**Source:** UNICEF, 1994.

**Table 8.2: Public spending per pupil on primary and secondary education (US$) by region, 1980 and 1990**

| Region | 1980 | 1990 |
|---|---|---|
| Sub-Saharan Africa | 62 | 58 |
| East Asia Pacific | 32 | 76 |
| South Asia | 62 | 104 |
| Arab States | 179 | 263 |
| Latin America/Caribbean | 165 | 267 |
| Industrialized Nations | 1 327 | 2 419 |

**Source:** UNESCO, 1993.

Public expenditure as a percentage of Gross National Product (GNP) is a crude indicator of the priority placed on education (Lewin, 1993). Africa spends the most on education of any region in the world as a percentage of GNP but the least per pupil in absolute terms. In 1980 (see table 8.2), schools in sub-Saharan Africa and South Asia spent roughly the same amount per pupil. By 1990, however, spending per pupil in South Asia had increased by almost 70 % and fallen by almost 7 % in Africa. In Africa, increasing enrolment has not attracted a corresponding increase in physical, human and financial resources. Many African countries are struggling with the problem of an unstable, underqualified teaching force, particularly for science, mathematics and technology (Stoll, 1993).

Low government spending on education means poorly paid and poorly supported teachers, often working in dilapidated classrooms and laboratories with insufficient furniture and space. In overcrowded classes, choral recitation becomes the dominant mode of instruction.

In this context, the issue of class size and its relationship to outcome measures, such as pupil achievement and teacher satisfaction, has become a political as well as a professional issue (Smith & Glass, 1979). Policy makers and donors have come to demand that increased expenditure be justified by a corresponding increase in pupil achievement. However, some educational decision makers and stakeholders believe that large classes have compromised the quality of education beyond acceptable levels. Thus research is needed to establish what effects class size has on the quality of teaching and learning and whether pupil achievement can be improved under current constraints. Since large classes will be a reality in Africa for the foreseeable future, teacher educators must find ways to overcome the inherent infrastructural and management constraints they place on good teaching.

This chapter addresses some of the problems that teaching large classes poses for effective science and technology education. It outlines characteristic features of the prevailing classroom environment, considers what past research has to say about the relationship of class size to quality of learning, and suggests some strategies for teaching large classes that maximize pupil involvement and resource utilization. The discussion focuses on the secondary level of education, where large class size may be a particularly important problem in science, technology and mathematics teaching. Finally, the paper makes a plea for more action research into effective ways to teach inquiry science to large classes that have few resources.

## RESEARCH ON CLASS SIZE

There are other constraints than class size that must be addressed if teachers are to change their views of science as well as their teaching methods. An example is the memorization-oriented examinations that force teachers to view the science curriculum narrowly as facts and concepts in a syllabus that they must cover. A recent World Bank publication, *Priorities and Strategies for Education* (World Bank, 1995: 101), urges that a wide variety of performance indicators should be used in addition to examinations. Arguably such constraints may be more important than large classes or limited standard equipment, and until they are removed, little will change. What does research have to say on the specific constraints posed by large class size? Is there a relationship between large classes and outcome measures such as pupil achievement and teacher satisfaction?

### Class size and classroom outcome measures: What does the research say?

Decreasing class size is the most controversial technique that has been proposed to improve the quality of education (Smith & Glass, 1979; Walberg, 1991). There are conflicting arguments for and against reducing class size. The literature cuts across levels of education and subject disciplines.

Teachers swear by the benefits of small classes. Policy makers and administrators, on the other hand, focus on the higher costs involved, demanding that smaller classes be justified on the basis of increased pupil achievement (Smith & Glass, 1979).

Research has been unable to resolve the controversy. Some studies show that pupils do better in smaller classes (Glass & Smith, 1978, 1979); some suggest that large classes are more effective, given appropriate teaching methods (Moock & Harbison, 1987; Hanushek 1986) and many fail to reach a conclusion.

Hanushek (1986) concluded that '. . . the available evidence in more than 150 studies suggests no relationship between expenditures and pupil achievement, attitudes, and dropout rates, [and] traditional remedies such as reducing class size or hiring better-trained teachers are unlikely to improve the matter'.

Empirical studies of the World Bank and other agencies show a range of factors that affect pupil achievement. For example, Fuller and Heyneman (1989) ranked eight factors that have positive effects on learning in Third World countries. These are:

 

| **Highly effective factors** | % |
|---|---|
| Length of instructional programme | 86 |
| Pupil feeding programme | 83 |
| School library activity | 71 |
| Textbooks and instructional materials | 67 |
| **Less effective factors** | |
| Science laboratories | 36 |
| Teacher salaries | 36 |
| Reduced class size | 24 |
| Pupil grade repetition | 20 |

Haddad (1979) was unable to find consistent evidence that reduced pupil–teacher ratios and smaller class sizes improve educational quality in developing countries.

In discussing secondary education in sub-Saharan Africa, Moock and Harbison (1987) recommended both incremental and radical improvement in educational productivity. They suggested that class size in secondary schools might be increased substantially without sacrificing quality.

The results from such broad reviews must be viewed cautiously. They do not establish cause and effect or distinguish the role of local circumstances. Indeed, class size and science laboratories may be irrelevant to pupil performance in rote-memory examinations but may affect performance in examinations that assess higher-order thinking skills. Class size may affect pupil achievement through intervening variables, such as teacher support services, pupil attitudes and classroom environment.

In an exploratory study of Nigerian schools, Alonge (1985) investigated the effect of class size on the achievement of chemistry pupils in various ability groups. Preliminary findings showed no significant differences in performance between pupils in classes of 40, 60 or 120. However, a related study (Ndukwe, 1995) compared the achievements of senior secondary pupils in laboratory classes of 30 and 100. Pupils in the smaller classes performed significantly better. Performance differences were attributed to a shortage of instructional materials and facilities in the larger classes and the inability of teachers to respond to individual needs.

Japan commonly has classes of 40 to 60 but surpasses nearly all Western countries on standardized tests of secondary-school-level mathematics, and science knowledge and comprehension (Walberg, 1991). Walberg goes on to suggest possible reasons for the high performance of Japanese students in science. Teachers '... ask hard, provocative questions; entertain many thoughtful pupil answers while suspending judgment; elicit decisive designs for experiments from pupil teams; and still suspending judgment, allow pupils to take the lead designing, conducting, and interpreting the experiments done with simple everyday equipment and materials' (Walberg, 1991: 48). Such methods ensure a high level of pupil involvement. By contrast, other studies indicate that reduced class size can have positive effects on pupil learning. Glass and Smith (1978, 1979) examined the relationship between class size and pupil achievement through a statistical integration of 80 existing studies. They demonstrated a substantial relationship between class size and achievement. Their results showed that:

> As class-size increases, achievement decreases. A pupil, who would score at about the 63rd percentile on a national test when taught individually, would score at about the 37th percentile in a class of 40 pupils. The difference in being taught in a class of 20 versus a class of 40 is an advantage of 10 percentile ranks . . . Few resources at the command of educators will reliably produce effects of that magnitude. (Glass & Smith, 1978: 1)

Smith and Glass (1979) also reviewed research results on the relationship between class size and classroom transactions, teacher satisfaction and effect on pupils. Again, their results showed a positive impact of reduced class size.

- Reducing class size has beneficial effects on cognitive and affective outcomes, and on the teaching process.
- Class size affects the quality of the classroom environment. In a smaller class there are more opportunities to adapt learning programmes to the needs of the individual. Pupils are more directly and personally involved in learning.
- Class size affects pupils' attitudes either as a function of better performance or contributing to it.
- Class size affects teachers. In smaller classes their morale is better; they like their pupils better, have time to plan and are more satisfied with their performance.
- On all measures, reduction in class size is associated with better schooling and more positive attitudes.
- Class size effects were related to the age of pupils, with effects most notable for those 12 years and under and least apparent for pupils over 18.

As the authors point out, improved academic achievement is not the only justification for class size reduction. Moreover, the notion that in a small class there are more opportunities for teachers to innovate and adapt learning programmes to the needs of individual pupils does not necessarily mean that teachers will do so. Some will need training. Others will continue to use traditional methods even with a class of five.

Unfortunately, because the research evidence appears to be conflicting, the debate over class size has become highly politicized. In Africa, however, large classes are the reality. Thus the argument is not whether large classes are good but whether there are suitable ways to promote good teaching in large classes.

Teachers in African countries know that it is difficult to work with large numbers of pupils. Nevertheless, some achieve a high level of pupil involvement. How can all science teachers faced with large classes maintain a conducive learning environment that promotes pupil activity, that provides opportunities for first-hand experiences, that challenges pupils to ask questions and initiate the learning process, and uses local resources to overcome infrastructural constraints?

## THE REALITY OF LARGE CLASSES

In Nigeria, recommended teacher–pupil ratios are 1: 30 for primary and 1: 35 for secondary schools. In reality, in some states they are between 1: 50 and 1: 85 (Ajewole, 1995). This paper defines a large class as one that has a teacher–pupil ratio larger than 1: 40, and an overcrowded class as one where available floor space is less than one square metre per child. In much of Africa, not only are classes large but, unlike large classes in the industrialized world, they are frequently also overcrowded, and lack resources.

The prevailing culture of science and technology teaching throughout Africa is one of imparting predetermined and highly structured knowledge. The syllabuses and recommended textbooks reinforce the dominant position of the teacher and the textbook as major sources of information, with pupils as passive recipients of knowledge. Within this environment, pupil initiative is stifled and interaction between pupils becomes abnormal.

Yet the objectives of the new science and technology curricula call for a classroom environment that encourages pupil activity, provides first-hand experiences that challenge pupils to initiate learning, uses their existing ideas to make sense of the content to be learned, and engages them in practical work. By contrast, pupils are taught by lectures, the blackboard and occasionally through demonstration of standard experiments.

This contradiction must be faced. Either we must accept the impossibility of hands-on inquiry learning in our science classes and lower the expectations we have of teachers, or we must evolve courses, teaching materials and approaches that make inquiry possible within the realities of African classrooms. I believe a compromise is possible. A recent nationwide survey in Nigeria (Yoloye, 1989) identified the most prevalent approaches to science teaching in order of frequency as: (1) teaching or explaining new content to the entire class; (2) revising old content with the entire class; (3) whole-class discussion; and (4) demonstration of experiments by the teacher. In about 40 % of the schools surveyed, teachers spent significant amounts of time maintaining discipline.

During a recent workshop, science teacher educators in the Science and Mathematics Education Unit of the Department of Teacher Education, University of Ibadan, conducted a survey of primary and secondary science teachers to determine their views on teaching large classes. Each participant was asked to respond briefly in writing. The teachers' views on teaching large classes included the following:

When teaching large classes:

- It is more difficult for teachers to do practical work and to give pupils individual attention.
- Pupils learn less because most of them are not actively involved and become easily distracted.
- Teacher-centred teaching is encouraged, restricting the range of teaching and assessment strategies and making it difficult to identify pupils' learning difficulties or needs.

- Frequent practical work becomes difficult.
- The level of pupil participation is low.
- There are heavier demands on facilities and instructional materials.
- There is minimal class control and supervision and because of this pupils learn less.
- Since science is an experimental subject — without hands-on activities (which large classes do not allow) — one cannot be said to be doing science.
- Movement and laboratory activities become restricted.
- It is difficult to provide a classroom environment conducive to learning.
- There are opportunities for peer-tutoring that can be effective.
- There are opportunities for cooperative group work.

We also surveyed pupils' views on large classes. They reported that in large classes:

- There is no climate for sustained concentration, and as a result there is more apathy and frustration.
- Teachers frequently do not give follow-up assignments because of the workload involved in marking.
- Only a few pupils actively participate, because teaching and learning resources are limited.
- There is little or no individual practical activity and most experiments are conducted as teacher demonstrations.
- Less academically motivated pupils are left to their own devices since they 'hide' in the anonymity of the crowd and have little interest in learning.
- There is little or no opportunity to handle apparatus and equipment.
- Teaching methods are predominantly 'chalk and talk'.
- Discipline is difficult to maintain.
- Distractions affect pupils' attitude to work.
- There are few opportunities for pupils to work in groups.

It became apparent that the problems identified relate to a view of science as 'something only done in classrooms' or as 'something only scientists do', using the 'tools' of science.

Primary teachers surveyed considered it important to expose their pupils to the processes of science, while secondary teachers thought they should emphasize concepts. However, large class size discourages pupils who wish to engage in independent inquiry; instead, experimental work is restricted to an occasional teacher demonstration. This lack of access to inquiry contributes to the perception of science as 'just something that we do in schools or labs' (Nesbitt, 1993).

We must help teachers to de-emphasize the image of science as 'something done only in classrooms' or as 'something only scientists do', and instead emphasize an image that places science firmly in the learners' world.

> Science must become relevant and accessible to pupils of differing abilities, interests and culture. It must be based on situations they encounter in their local environment and build upon their creativity, curiosity and existing knowledge (UNESCO, 1989).

Such science should cause teachers to redefine their thinking about the nature of science and make it consistent with individual and societal needs (Nesbitt, 1993). It would help pupils and teachers to perceive and experience science as a human activity. It also has important implications for teaching methods used in science classrooms.

To scientists and technologists there is little distinction between theory and practice. Theory provides a basis for experimentation that in turn modifies theory. In this sense, conventional science teaching is neither theoretical nor practical — regardless of class size or provision of resources. It is, rather, based on memorization (theory) followed by prescribed experimental procedures that, if correctly followed, provide the expected results (practicals). Such teaching undermines children's curiosity and minimizes opportunities for active participation in theory building through experimentation. There must be structure, but a structure based on pupil involvement in the problem-solving processes of science.

## TEACHING LARGE CLASSES: ISSUES TO BE ADDRESSED

Though large classes are created by the system, coping is a management issue for individual teachers. Policy makers and curriculum developers must seek ways to support them while teachers and teacher educators must develop effective ways to teach large classes.

Teaching methods carry empowering and disempowering messages. An empowering teacher uses strategies that encourage pupils to question nature and to investigate problems. Such teachers encourage pupils to extend their interest and experience beyond the classroom and the textbook. Teachers must be assured that their methods do empower pupils (Onwu, 1992) and enable them to become responsible for their learning.

The effects of class size on teacher satisfaction are strong and there are a number of different ways to offer teachers support. Basically the issues are how teachers of large classes can be helped to: (1) adopt strategies that provide for more pupil involvement; (2) use classroom management techniques that maximize resource utilization; (3) recognize local resources; and (4) relate local resources to topics in their curriculum.

### Classroom management for pupil involvement

When suggesting ways to encourage greater pupil involvement in large classes we should consider the available equipment, timetable constraints, costs, pupil attitudes and age, the language of science instruction, teachers' views of science, social views of learning and the expectations of society. Our recommendations should be predicated on:

- A classroom shift from teaching to learning, so that pupils are encouraged to become more responsible for their own learning. Teachers should facilitate learning rather than be the predominant source of all information.

- Encouraging pupils to share ideas and ask questions of each other and of the teacher. Pupils should appreciate that their ideas are important and should develop a concept of themselves as both teachers and learners. Pupils should initiate scientific investigations and cooperative approaches to learning.
- Not expecting teachers to bear the full burden of change. Appropriate curriculum materials, syllabuses, examinations, inspectorate behaviours and so on must be in place.
- Teaching methods that relate to the sociocultural context of the learner.

The following suggestions are for teachers to workshop and trial.

### *Asking questions*

Pupils should be encouraged to express their ideas. They should ask questions of each other and of teachers. This may be threatening if pupils lack confidence or are shy. Cultural factors may stifle creativity, initiative and the asking of questions (Onwu, 1990). One suggestion is to ask pupils to write down their ideas and questions.

### *Working in groups*

Working in small groups during practicals can engage pupils in doing science and encourages pupil-centred learning. If pupils work on different problems, scarce learning materials become more accessible. However, pupils' skills of working in groups need to be developed since they often have limited experience of doing so, having been exposed mostly to didactic teaching. Research shows that the process of allocating pupils to groups is critical (Damon, 1984; Okebukola, 1986; Onwu & Ojo, in press). Allowing pupils to decide with whom they wish to work and ensuring individual accountability appears to increase pupils' involvement, though in practice, the allocation is usually done arbitrarily by teachers (Naidoo & Reddy, 1994).

### *Cooperative learning*

Cooperative learning seems a fruitful way to teach large classes (Johnson & Johnson, 1983; Webb, 1984). It involves delegating some control of pacing and methods of learning to pupil groups of 3 to 6 members, who work together, sometimes competing with other groups (LaCombe, 1992). A cooperative learning environment is fostered through shared goals and the accountability of each group member (Okebukola & Ogunniyi, 1984; Sapon-Shevin & Schiendewind, 1992). Pupils work together on assigned tasks, make decisions by consensus and ensure that each member contributes. Data from a class of 81 senior secondary biology pupils in Nigeria confirmed that cooperative learning is an effective way to promote pupil achievement (Okebukola, 1984). It goes beyond group work (LaCombe, 1992; Naidoo & Reddy, 1994) and facilitates problem solving by learners, the pooling of information, effective discussion, individual contribution and peer-tutoring (Onwu & Ojo, in press). It also better promotes positive attitudes and problem-solving competence than individualistic learning. Pupils' discussion and communication skills improve as they ask questions, initiate actions and provide ideas. Letting pupils select whom they wish to work with, identify problems, and discuss how they will investigate, maximizes their involvement. Naidoo and Reddy (1994) note that cooperative learning became successful in South Africa only as a result of classroom-based action research. However, teachers' lack of action research skills and of organizing cooperative learning, together with prescribed curricula and examinations, are limiting factors to the adoption of cooperative learning.

### Pre-service education for pupil involvement

Pre-service teacher education programmes should include imparting the skills of organizing cooperative learning and other strategies that widen teachers' understanding of the term 'curriculum'. Student teachers' experience should move them from regarding textbooks as the curriculum to using action research in the classroom. College training should involve student teachers in cooperative learning as well as exposing them to model classrooms or video tapes. Students could then analyse the strategy, discuss it and try it in schools. Professional teaching associations could encourage cooperative learning.

## LARGE CLASSES AND RESOURCE UTILIZATION

There are many resources in the African environment for teaching science and technology, and the prevailing economic realities make it imperative that we increase teachers' awareness of them. They would include people, industry and other institutions, materials, media, local technologies, culture and the natural environment.

### Recognizing the local environment as a resource for teaching

Pupils can be ingenious in identifying and collecting materials. They suggest resources for activities in which they are interested and explore questions about things with which they are familiar. Such pupil interest can provide a starting point for teachers to extend learning.

However, many teachers still cannot link local resources, environments and cultures to topics in the curriculum and recommended textbooks often provide little assistance. Though many African countries provide appropriate books and equipment for teaching primary school science (UNESCO, 1988), many secondary school science teachers need help to see how everyday materials can provide a basis for learning. The STAN journal regularly features suggestions for teachers on how to use local materials and how to carry out task analyses of syllabus topics.

Teachers could use equipment to pose problems rather than to demonstrate concepts. They could ask classes to use demonstration equipment to explore solutions. Though only selected pupils would handle the apparatus, the whole class could become involved in a 'minds-on' activity. Working this way has the advantage that no new knowledge is expected from teachers. The approach simply asks teachers to become involved in problem solving rather than merely demonstrating solutions.

Teachers could involve students in 'thought experiments', using the chalkboard, drawings, photographs and so on to pose problems. Pupils in groups could decide what they would need and what procedures they would use to solve the problems. Should teachers have some equipment, a few pupils could be asked in order to demonstrate their solutions. Their suggestions would lead other pupils to proceed further in their imaginary investigation. This approach is similar to that used by the British primary school project, 'Think and Do', that is being modified by Professor James Otuka for use in Nigerian schools.

The training and support of teachers at local levels is necessary to sensitize them to resources available in their local areas and to their effective use. This could be done by:

- Holding workshops where teachers locate and describe local resources, discuss how they could be used and share their ideas.
- Helping groups of teachers and science educators to prepare tables that link curriculum topics to specific local resources and appropriate teaching activities.
- Encouraging industries and tertiary educational institutions to publish local materials for schools.

At the national level, curriculum guides could describe what equipment pupils can make from local resources and how teachers can use local technologies and personnel.

For instance, Science and Technology in Action in Ghana (STAG), a project of the University of Cape Coast, sponsored by the African Forum for Children's Literacy in Science and Technology (AFCLIST), has produced a resource book that describes technologies in Ghana. Learning materials are based on the science inherent in local industry and manufacturing. Multimedia packages for in-service and pre-service science teacher education are planned.

The production and dissemination of such packages would help teachers identify local resources they can use to develop their teaching programmes, rather than starting from abstract concepts. By starting with local phenomena, the same concepts could be developed, the national curriculum fulfilled, and pupil learning would become more interesting and meaningful.

Teachers need a knowledge of pedagogy as well as of content. Pre-service and in-service teacher programmes should help teachers acquire confidence in using different teaching methods (Onwu, 1985). In teaching poorly resourced, large classes, initiative must be shown (Onwu, 1985). Methodology courses should prepare teachers to vary their teaching approaches and to use a combination of discovery and expository methods that include teacher demonstration, pupil experimentation and project work to develop pupils' scientific knowledge, skills and attitudes.

## CONCLUSION

Large, overcrowded and poorly resourced classes are a reality in most African countries, a reality that science educators must face. Because of the health hazards of overcrowded classrooms, policy makers must make it an immediate priority to reduce their sizes. For science educators the priority is action research into ways of teaching inquiry science effectively to large classes.

## REFERENCES


Ajewole, GA. 1995. Large class and practical works in Science Technology and Mathematics (STM): An investigation into the effects of guided inquiry learning approach. Paper presented at the 36th Annual STAN Conference, 14–19 August, Maiduguri, Nigeria

Ajeyalemi, D, Collinson, G, Aidoo Taylor, N, Middleton, P, Baiyelo, S, Imenda, N, Ibikunle, Johnson V, Musonda, D, Hodzi, R & Chagwedera, M. 1990. Science and technology education in Africa. In Ajeyalemi, D (ed). *Focus on Seven Sub-Saharan Countries.* Lagos: University of Lagos Press

Alonge, EI. 1985. Teaching chemistry to large classes: An exploratory study. *Journal of Research in Curriculum,* 3(2)

Bajah, ST. 1995. Education and politics: Their interplay as stabilising factors. Paper presented at the Nigerian Educational Conference on Stabilising the Nigerian Educational System, Federal College of Education, April 26–28 Abeokuta, Nigeria

Damon, W. 1984. Peer education: The untapped potential. *Journal of Applied Developmental Psychology,* 5

Fuller, B & Heyneman, ST. 1989. Third World school quality: Current collapse, future potential. *Educational Researcher,* 8(2)

Glass, GV & Smith, ML. 1978. *Meta-Analysis of Research on the Relationship of Class Size and Achievement.* San Francisco: Far West Laboratory for Educational Research and Development (No Ob-NIE-G-78-0103, Dr LS Cahen, Project Director)

Glass, GV & Smith, ML. 1979. Meta-analysis of the research on class size and achievement. *Educational Evaluation and Policy Analysis,* 1

Haddad, WD. 1979. Educational Effects of Class Size. Working Paper No 280. Washington, DC: World Bank

Hanushek, EA. 1986. The economics of schooling: Production and efficiency in public schools. *Journal of Economic Literature,* 14

Johnson, DW & Johnson, RT. 1983. The socialization and achievement crisis. Are cooperative learning experiences the solution? In Bickman, L (ed). *Applied Sociology Annual,* vol 4(2). Beverly Hills: Sage Publications

LaCombe, S. 1992. Editorial comment: Cooperative learning at a crossroads. *Journal of Education,* 174(2)

Lewin, K. 1993. Planning policy on science education in developing countries. *International Journal of Science Education,* 15(1)

Moock, PR & Harbison, RW. 1987. *Education Policies for Sub-Saharan Africa: Adjustment, Revitalization and Expansion.* Washington, DC: World Bank, Population and Human Resources Department

Naidoo, P & Reddy, S. 1994. Teaching of a large science class: A case study at University of Durban-Westville, South Africa. Paper presented at NARST Annual Meeting, 26–30 March, Anaheim, California

Ndukwe, KI. 1995. Impact of small group and large group practical classes in pupils' performance. Paper presented at the 36th Annual STAN Conference, 14–19 August, Maiduguri, Nigeria


 

Nesbitt, JE. 1993. Teaching science in an artistic way. *Science Education International,* 4(3)

Okebukola, P. 1984. Teaching the problem large classes in biology: An investigation into the effect of a co-operative learning technique, *Journal of the Science Teachers' Association of Nigeria*, 22(2)

Okebukola, P & Ogunniyi, MB. 1984. Cooperative, competitive and individualistic interaction patterns: Effect on pupil achievement and acquisition of practical skills. *Journal of Research in Science Teaching,* 22(9)

Onwu, G. 1985. How should we educate teachers of senior secondary chemistry? *The College Review,* 3(1, 2)

Onwu, G. 1990. Development of creativity and of initiative in the context of African culture. *African Thoughts on the Prospects of Education for All.* Dakar: UNESCO-UNICEF

Onwu, G. 1992. Conducive classroom environment for science technology and mathematics: Implications for the learner. In Conference Proceedings, Science Teachers' Association of Nigeria 33rd Annual Conference, 17–22 August, Enugu, Nigeria

Onwu, G & Ojo, DA. In press. Differential effectiveness of cooperative competitive and individualistic goal structures on pupils' chemical problem solving performance. *Journal of the Science Teachers' Association of Nigeria,* 29(1, 2)

Sapon-Shevin, M & Schiendewind, N. 1992. If cooperative learning's the answer, what are the questions? *Journal of Education,* 174(2)

Schiller, DS & Walberg, HJ. 1982. Japan: The learning society. *Educational Leadership,* 39

Smith, MC & Glass, GV. 1979. *Relationship of Class Size to Classroom Processes, Teacher Satisfaction and Pupil Affect: A Meta-Analysis.* San Francisco: Far West Laboratory for Education Research and Development

Stoll, CJ. 1993. Science education in developing countries: What's the point? *Science Education International,* 4(4)

UNESCO. 1993. World Education Report. Paris: UNESCO

UNESCO. 1989. *Science for All: Supporting Teacher Change.* Bangkok: UNESCO

UNESCO. 1988. *Innovations in Science and Technology Education*, vol 2. Paris: UNESCO

UNICEF. 1994. Education: Achievement and disparity. *The Progress of Nations.* New York: UNICEF

Walberg, HJ. 1991. Improving school science in advanced and developing countries. *Review of Educational Research,* 61(1)

Webb, NM. 1984. Sex differences in interaction and achievement in cooperative small groups. *Journal of Educational Psychology,* 76(1)

World Bank. 1995. Priorities and Strategies for Education. Washington, DC: World Bank

Yoloye, EA. 1989. *Survey of Resources in Science, Mathematics and Technical Education in Secondary Schools in Nigeria.* Lagos: Federal Ministry of Science and Technology

# 9

# Resourcing science and technology education

*Emmanuel Fabiano, Chancellor College, Malawi*

## ABSTRACT

The success or failure of science and technology education is dependent on the availability and utilization of appropriate resources. This chapter focuses on the quality and quantity of teachers; the role and use of print and learning materials; the impact of laboratory space, equipment and consumables on the effectiveness of practical work; the use of the school environment; and financial resources. It discusses the question, can Africa resource science and technology education on a self-sustaining basis?

## INTRODUCTION

Education, like industry, has definable products. Since product quality is a function of the inputs and the processes that convert them, it is important that planners analyse these factors carefully to identify sources of weakness. On the basis of such understanding, policy makers would be better able to make decisions on how to improve product quality.

Mjojo (1994) documented the importance of science and technology to development. Since attaining independence, the commitment of African states to science and technology education has been striking. For example, communiqués of successive conferences of African ministers of education (Addis Ababa, 1961; Tananarive, 1962; CASAFRICA 1, 1974; and the Lagos Plan of Action, 1981) contain strong statements of support. African governments allocate significant percentages of gross national products (GNPs) to education; more children, including girls, are in school for longer periods; and they are being taught more science (see table 9.1). Countries such as Nigeria have ratios closely approaching 60 : 40 of students at secondary schools and universities studying science, compared with those studying the humanities (Ivowi, 1995). Some countries have proceeded through several generations of curriculum development in science and technology at all levels of education (Caillods, Gottelmann-Duret & Lewin, 1995).

However, such investment has not led to the anticipated results. Odhiambo (1993) estimates that, despite these impressive efforts, most African countries have fewer than three scientists and engineers per 1 000 graduates and that the industrial sector employs only 7 % of the workforce. Though governments spend large percentages of recurrent budgets on education, expansion has led to funds being spread thinly. High population growth and enrolment rates have led to a decrease in per capita spending over the past decade (table 9.1). Since teachers' salaries consume the largest percentage of the education budget, fewer funds are now available for books, equipment and support services and this contributes to deteriorating standards (table 9.2). Expansion of enrolment rates at secondary and tertiary levels, with concomitant increases in per capita student expenditures (table 9.1), exacerbates the situation, yet Africa continues to have the lowest enrolment rates at these levels of any region in the world. The economic crises experienced by many countries in Africa, together with competing demands from other sectors such as agriculture and health, make it unlikely that education budgets will grow in the foreseeable future. Nor can much be expected from donors such as the World Bank, since despite their influence on education systems in Africa, their contribution amounts to only about 2 % of education budget. The challenge for Africa is how to provide equitable, quality education that includes science and technology education, with little extra government funding.

**Table 9.1: Summary statistics on educational development and financing**

| Country group | Pop growth 1990–93 | % Pop 6–14 years | GNP/ Cap US$ 1990 | GNP growth 1980–90 | Gross enrolment rate (Pri) | Gross enrolment rate (Sec) | Gross enrolment rate (Tert) |
|---|---|---|---|---|---|---|---|
| **GNP/Cap < US$1 000** | | | | | | | |
| Average (61) | 2,7 | 44,3 | 454,6 | 2,1 | 79,0 | 25,7 | 4,7 |
| Average SSA (36) | 3,0 | 47,2 | 376,1 | 1,7 | 72,0 | 17,1 | 1,9 |
| Average Anglo Af (15) | 3,1 | 47,3 | 355,6 | 1,2 | 79,3 | 26,8 | 3,3 |
| Average Franco Af (17) | 3,0 | 46,8 | 417,1 | 1,8 | 66,8 | 13,5 | 1,8 |
| Average Luso Af (4) | 2,6 | 46,5 | 340,0 | 3,2 | 72,5 | 11,3 | 0,5 |
| Average S–E Asia (11) | 2,4 | 38,1 | 369,1 | 3,6 | 68,4 | 38,1 | 6,6 |
| Average C Am/Carib (6) | 2,4 | 41,7 | 578,3 | 1,8 | 60,5 | 15,7 | 2,9 |
| Average Ger <90 (Pri) | 2,8 | 45,7 | 360,3 | 1,8 | 60,5 | 15,7 | 2,9 |
| Average Ger >90 (Pri) | 2,6 | 42,5 | 565,7 | 2,4 | 105,7 | 40,6 | 6,7 |
| **GNP/Cap US$1 000–5 000** | | | | | | | |
| Average (44) | 1,7 | 32,9 | 2 203,2 | 2,4 | 105,3 | 58,3 | 15,1 |
| Average SSA (7) | 2,5 | 38,5 | 2 938,6 | 3,0 | 114,0 | 45,7 | 3,5 |

| Average S Am (8) | 1,9 | 31,8 | 1 952,5 | 1,4 | 108,4 | 56,1 | 23,0 |
|---|---|---|---|---|---|---|---|
| Average C Am/Carib (14) | 1,4 | 34,6 | 2 345,0 | 3,3 | 104,4 | 54,6 | 14,0 |
| Average Eur (9) (6) | 0,3 | 19,8 | 2 726,7 | 1,6 | 97,8 | 80,0 | 19,1 |
| Average Asian NICS (5) | 1,5 | 24,4 | 6 602,0 | 5,6 | 101,2 | 64,2 | 12,1 |
| **GNP/Cap > US$5 000** | | | | | | | |
| Average | 1,4 | 22,2 | 15 378,9 | 3,4 | 102,0 | 88,7 | 30,1 |
| Average Eur/N Am (21) | 0,6 | 18,2 | 17 946,7 | 2,2 | 101,7 | 95,2 | 34,7 |
| Average Gulf (7) | 3,8 | 34,0 | 11 495,7 | 5,6 | 99,0 | 72,2 | 17,2 |

| **Country group** | **Teacher–pupil ratio (Pri)** | **Teacher–pupil ratio (Sec)** | **% GNP on education 1990** | **% Govt exp on education 1990** | **Growth in educ exp 1980–90** | **Exp/pupil as % GNP per capita** | | |
|---|---|---|---|---|---|---|---|---|
| | | | | | | **Level 1** | **Level 2** | **Level 3** |
| **GNP/Cap < US$1 000** | | | | | | | | |
| Average (61) | 39,7 | 21,8 | 3,9 | 15,9 | 5,0 | 0,11 | 0,43 | 4,59 |
| Average Ger > 90 (Pri) | 35,4 | 21,2 | 4,6 | 14,6 | 5,2 | 0,10 | 0,27 | 1,94 |
| Average SSA (36) | 44,3 | 23,6 | 4,0 | 16,3 | 4,3 | 0,13 | 0,53 | 7,01 |
| Average Anglo Af (15) | 36,8 | 21,8 | 4,8 | 15,1 | 5,8 | 0,12 | 0,58 | 6,03 |
| Average Franco Af (17) | 52,3 | 24,1 | 3,3 | 18,3 | 3,1 | 0,13 | 0,48 | 6,05 |
| Average Luso Af (4) | 37,0 | 34,0 | 4,6 | 11,4 | 0,1 | 0,23 | 0,56 | 22,24 |
| Average S–E Asia (11) | 36,2 | 20,2 | 3,4 | 10,4 | 7,9 | 0,08 | 0,20 | 0,89 |
| Average C Am/Carib (6) | 34,8 | 19,7 | 3,1 | 15,9 | 5,7 | 0,11 | 0,16 | 1,24 |
| Average Ger < 90 (Pri) | 42,9 | 22,2 | 3,3 | 17,0 | 4,8 | 0,13 | 0,54 | 6,71 |
| **GNP/Cap US$1 000–5 000** | | | | | | | | |
| Average (44) | 25,9 | 17,6 | 5,2 | 15,2 | 3,7 | 0,11 | 0,20 | 0,87 |
| Average SSA (7) | 36,4 | 19,3 | 6,1 | 13,5 | 5,9 | 0,06 | 0,41 | 1,68 |
| Average S Am (8) | 24,9 | 14,7 | 3,7 | 18,7 | 3,4 | 0,07 | 0,10 | 0,47 |
| Average C Am/Carib (14) | 27,2 | 19,1 | 5,1 | 14,0 | 2,4 | 0,12 | 0,16 | 0,01 |
| Average Eur (9) (6) | 15,5 | 14,7 | 4,9 | 9,5 | 2,9 | 0,16 | 0,18 | 0,55 |
| Average Asian NICS (5) | 25,0 | 21,4 | 4,1 | 15,7 | 7,3 | 0,10 | 0,10 | 0,40 |
| **GNP/Cap > US$5 000** | | | | | | | | |
| Average | 18,2 | 14,2 | 5,2 | 13,8 | 3,5 | 0,16 | 0,22 | 0,41 |
| Average Eur/N Am (21) | 16,6 | 12,6 | 5,5 | 12,8 | 2,9 | 0,17 | 0,23 | 0,37 |
| Average Gulf (7) | 17,9 | 12,1 | 4,3 | 13,5 | 3,6 | 0,10 | 0,24 | 0,50 |

**Table 9.2: A statistical profile of education in sub-Saharan Africa in the 1980s**

| Country | Primary | | Secondary | | Teacher | Tertiary | |
|---|---|---|---|---|---|---|---|
| | Total | Teaching materials | Total | Teaching materials | Training | Total | Teaching materials |
| Ethiopia | 51 | 0,4 | 77 | 1,1 | 817 | 851 | – |
| Malawi | 50 | 0,3 | 192 | 23,2 | – | 1 782 | 16 |
| Tanzania | 12 | 2,6 | 132 | 5,1 | 401 | 1 412 | 166 |
| Rwanda | 54 | 1,1 | 182 | 3,6 | – | 4 050 | 29 |
| Botswana | 219 | 5,3 | 1 302 | 86,7 | 7 392 | 7 218 | – |
| Sub-Saharan Africa (Median) | 49 | 1,7 | 192 | 13,5 | 558 | 1 971 | 40 |
| The Gambia | 42 | 2,7 | 98 | 2,5 | – | – | – |

**Source:** UNESCO, 1994b.

Education is generally accepted as an instrument of change (Hallak, 1990). For the first time, the World Bank (1996) has factored people as well as natural resources and capital assets as components contributing to individual and national wealth, as well as strengthening civil institutions and thereby good governance. Investment in the right sort of science and technology education does have an impact. However, factors other than financing may be equally significant. Such factors may include the influence of local cultures on learning, an inability to exploit available resources (UNECA, 1994) or, indeed, the type of learning to which we expose our students. Although economic development takes place within a complex web of interrelated factors, science and technology educators bear a responsibility to review past experience and examine the available options to make learning more effective.

Fabiano (1980) argues that effective teaching and learning depend on the learner and available resources. Such resources include the teacher, print materials, laboratory space, equipment and consumable supplies, the school environment, students, and funds. I shall be discussing these with a view to assessing their strengths, weaknesses and potentials in order to make recommendations for the future.

## DISCUSSION

Before we proceed with the discussion of resourcing, there are a number of questions that we should ask. For instance, what is the resourcing for, and what do we expect of learners as a result of their exposure to science and technology education? Does Africa need students who perform well on achievement tests by memorizing selected concepts and information? Or do African societies require problem solvers who can apply their learning, be they farmers or research scientists, as Makhurane argues in chapter 2 and Savage in chapter 3? Do countries in Africa need science and technology courses that emphasize content or processes? Do we need subject courses that stress science content, or integrated courses that include a consideration of the social implications of science and technology? How will Africa provide quality science and technology education for all and produce the high-level manpower needed to solve basic development problems? Other chapters in this book discuss such issues at length. I propose that African countries require problem solvers to attend integrated inquiry science courses that stress the use of the local environment, with specialization postponed to the senior secondary school level. How we answer such questions radically affects our resourcing strategies.

Another critical issue is that of change itself. Radical change has its costs and repeated change is even more costly, since it causes disturbances that initially often make the situation worse (Lewin, 1995a). Incremental change is slower, but less costly and likely to be more effective. It is necessary to institute new approaches to educational planning so that the scarce resources available in Africa can produce maximum benefits from the educational system.

An important issue is that of technology education. Throughout this chapter science and technology are discussed as one — indeed, so are biology, integrated science, environmental science and mathematics. Such assumptions may be valid at primary and lower secondary school levels since learning is based on the local environment and, as students inquire, subject distinctions become blurred. Clearly, as they proceed to senior classes, specialization assumes more importance. Table 9.3 summarizes the discussion. It is also important to point out that many learning objectives of technology education overlap with those of science education (Caillods, Gottlemann-Duret & Lewin, 1995). Integrating technological concerns into the science curriculum will almost certainly be cheaper than offering technology as a separate subject. If science subject matter is to be useful beyond school, it should have some technological flavour.

## Laboratories, equipment and consumables

There is considerable debate concerning the importance of practical activities in science and technology learning. Authorities such as Caillods et al (1995) and Akyeampong and Anamuah-Mensah (1993) claim their contribution to student achievement is questionable; that laboratory costs can be 10 times those of normal classrooms; and costs for the maintenance and supply of consumables are significant. They therefore argue that practical activities are not cost-effective and should be kept to a minimum.

However, such studies measure achievement by pass rates on examinations that are rote-memory oriented and where practical components test little other than an ability to correctly follow procedures to achieve predicted results. Curriculum goals, even in the most traditional syllabuses, call for more. Science educators, planners and researchers should question the examinations as well as the importance of student activities in promoting an understanding of science.

**Table 9.3: Comparison of conventional and inquiry science teaching**

| | **Conventional science** | **Inquiry science** |
|---|---|---|
| Lab supplies | Elaborate and expensive<br>Often used for one-off experiments<br>Higher costs | Ordinary classrooms<br>Maximum use of local resources<br>Multiple use of equipment supplied<br>Lower costs |
| School environment | Ignored | Used to the maximum |
| Teacher behaviour | Imparts knowledge | Facilitates learning |
| Teacher–pupil ratio | May be high | Must be lower |
| Pre-service | Emphasis on knowledge<br>Long training periods<br>Higher costs | Learns how to teach<br>Shorter training<br>Lower costs |
| In-service | Emphasis on knowledge<br>Long training periods<br>Higher costs | Learns how to teach<br>Shorter training. More school-based<br>Lower costs |
| Print | Multiple copies of standard text<br>Higher costs | Fewer copies of largest selection of reference books<br>Lower costs |
| Syllabi, exams, etc | Little input required to maintain | Considerable input required for initial change |
| Students | Passive recipients | Active learners. Can assist in resourcing |

Lectures may be a suitable way to teach some concepts and principles. However, others are better understood through observing demonstrations or activity-oriented learning. A compromise may be possible that does not lead to rote learning at the expense of development of problem-solving skills. Suggestions would include:

1. The design of courses and examinations that review the relationship between theory and practice. Often in the reality of schools neither are what is understood nor practised by scientists and technologists who view them as a continuum to deepen understanding. Too often, theory in schools becomes cramming information and practicals are 'cookbook recipes' to demonstrate the correctness of memorized principles.

2. Alternatives to conventional laboratory-based work that involve students in 'minds-on' rather than 'hands-on' activities — such as thought experiments, student-centred demonstrations, simulations, posing and solving problems using photographs, drawings and video presentations — and a more extensive use of resources in local environments are less costly alternatives that teachers could use (Onwu, 1995).
3. A review of equipment provided to ensure that selection takes into account multiple usage, cost, their contribution to learning, and ease of maintenance and replacement.
4. A later introduction of specialization in science, thus reducing the need for elaborate laboratories countrywide, specialized teacher training, costly equipment and practical examinations.

If these suggestions were adopted, the contribution of practical activities could assume its central role in science and technology learning at less cost than providing for traditional laboratory work. Students would also experience more effective learning. Contributions would include savings resulting from a reduced need for expensive laboratories, equipment and consumables; more equitable access to science and technology learning; as well as a more effective learning experience, thus achieving national goals of science and technology education in a more cost-effective fashion.

A note is required concerning the provision of locally produced science kits. Many countries have resorted to doing this as an alternative to, or to supplement costly imported equipment (Ross & Lewin, 1992). In countries such as South Africa and Nigeria, with large markets and manufacturing infrastructures, these kits have proved economically viable. In others, such as Kenya and Malawi, kit production units require heavy subsidies and even then most schools cannot afford them (Fabiano, 1993). Often science curriculum developers have not been involved in the design of kits, so the kits do not meet the requirements of new approaches to science teaching, thus reinforcing traditional laboratory practice. Experience with donor projects that have supplied science kits to schools in Ghana and Zanzibar has shown that unless teachers are trained in their use, they often do not even open the kits.

However, regardless of the role practical work assumes in science and technology education, its effective implementation depends on the quality of the teachers. Caillods, Gottelmann-Duret and Lewin (1995) observe that often teachers plan and conduct pupil experiments without assessing their contribution to understanding. Such practices are a reflection of how teachers were themselves taught and trained. It is important, both on cost and professional grounds, that teachers carefully consider experimental work and make informed decisions on which experiments pupils can do, which teachers can do effectively as demonstrations, and which can be omitted without reducing the effectiveness of the learning process.

## The school environment

The immediate school environment and community are rich but often neglected resources for science and technology teaching. Ignoring them and ignoring the knowledge base of every community in Africa increases the costs of supporting learning, contributes to the image of science being divorced from life and only practised in special laboratories, and denies students an opportunity to bring their life experiences to learning. This is contrary to modern learning theories, and leads to generations of students who can apply what they have learned only in examination halls (Jegede, 1995). Deciding how best to use an experimental approach and making the best use of the local environment demands high levels of professionalism and underlines the need to invest strongly in teacher development.

## Quantity and quality of teachers

Table 9.4 shows the rapid expansion of education in Africa between 1970 and 1990. In part this has been due to training top and mid-level human resources in vacuums left by colonial powers, in part due to high population growth rates, and in part also to political pressures (Passi, 1990).

Nevertheless, high illiteracy levels and low industrial productivity still characterize African countries. As illustrated in table 9.5 (UNDP, 1994), economies remain primarily agricultural.

In attempting to combat such socioeconomic problems, rapid expansion of education has created its own problems that include supply and quality of teachers. Despite a corresponding expansion of teacher training, teacher–pupil ratios remain

**Table 9.4: Enrolment by level of education 1980-1990 (in thousands)**

| Country | Primary | | Secondary | | Tertiary | |
|---|---|---|---|---|---|---|
| | 1970 | 1990 | 1970 | 1990 | 1970 | 1990 |
| Ghana | 1 420 | 1 945 | 99 | 871 | 5,4 | 19,0 |
| Malawi | 363 | 1 461 | 11 | 31 | 2,0 | 6,7 |
| Tanzania | 856 | 3 379 | 45 | 161 | 2,0 | 6,7 |
| Nigeria | 3 516 | 13 609 | 357 | 2 908 | 22,0 | 370 |
| Zambia | 695 | 1 461 | 56 | 195 | 1,4 | 15,3 |
| Kenya | 1 428 | 5 932 | 136 | 643 | 7,8 | 33 |
| Zimbabwe | 736 | 2 116 | 50 | 672 | 5,0 | 49,4 |
| Swaziland | 69 | 116 | 8 | 42 | 0,2 | 3,2 |
| Botswana | 83 | 284 | 5 | 62 | 1,1* | 3,7 |
| South Africa | 3 691 | 6 949 | 541 | 2 804 | 73 | 277 |
| The Gambia | 17 | 80 | 5 | 20 | – | – |

* 1980 figure.

**Source:** UNESCO, 1994b.

high. However, data conflict and differ widely between and within countries. Quoting from UNESCO publications, Lewin (1995b) claims that in Anglophone Africa, ratios are about 1:44 at primary levels and 1:24 at secondary. Such data must be disaggregated since class sizes in urban schools are often larger than those in rural areas. Onwu (1995) reports ratios in Nigeria as being between 1:50 and 1:85. Table 9.5 gives a clear picture of the situation at primary level. Since in most African countries teachers' salaries consume the bulk of recurrent costs of schooling — in some cases over 90 % — it may be unrealistic to expect any significant change in the near future. Instead, ways must be sought that enable teachers to be more effective when working with large classes (Onwu, 1995).

**Table 9.5: Development indicators of some developing countries, 1990–1992**

| Country | Adult literacy rate (age 15+, %) | Primary teacher–pupil ratio | Secondary technical enrolment (as % of total secondary) | % of labour force in industry | % of labour in agriculture |
|---|---|---|---|---|---|
| Algeria | 61 | 28 | 7,0 | 33 | 18 |
| Angola | 43 | 32 | 5,9 | 10 | 73 |
| Bangladesh | 37 | 63 | 0,7 | 13 | 59 |
| Botswana | 75 | 32 | 4,6 | 11 | 28 |
| Brazil | 82 | 23 | – | 25 | 25 |
| Cuba | 95 | 13 | 32,0 | 29 | 24 |
| Egypt | 50 | 24 | 20,9 | 21 | 42 |
| Ethiopia | – | 30 | 0,5 | 2 | 88 |
| Ghana | 63 | 29 | 2,5 | 11 | 59 |
| Guyana | 97 | 34 | 3,4 | 26 | 27 |
| Kenya | 71 | 31 | 1,6 | 7 | 81 |
| Lesotho | – | 55 | 3,6 | 33 | 23 |
| Malawi | 40 | 64 | 2,4 | 5 | 87 |
| Malaysia | 80 | 20 | 2,2 | 28 | 26 |
| Mauritius | 80 | 21 | 1,4 | 30 | 16 |
| Nigeria | 52 | 39 | 3,9 | 7 | 48 |
| Swaziland | – | 33 | 1,4 | 9 | 74 |
| Tanzania | – | 35 | – | 5 | 85 |
| Zambia | 75 | 44 | 2,8 | 8 | 38 |
| Zimbabwe | 69 | 36 | 1,7 | 8 | 71 |

**Source:** UNDP, 1994.

Apart from large teaching loads, many secondary school teachers lack laboratory assistants. Where they do have them, often the assistants are inadequately trained, so science teachers spend more time preparing experiments, further increasing their workload. As Lewin (1995b) points out, if trained assistants are paid 20 % of a teacher's salary and provided in a ratio of 1:5 teachers, this would increase the cost per pupil by 4 %. However, should doing so enable teachers to teach two extra periods a week, there would be a net gain in productivity of 1 % and possibly an improvement of the quality of experience offered to students.

Throughout Africa many secondary schools lack well-trained teachers. Often teachers who have never studied science during their training are forced to teach it (Yoloye, 1989). Some may even have failed science at O level. It is therefore not surprising that many students develop negative attitudes towards science during their secondary school education.

The situation is worse at primary school levels where teachers often teach all the subjects on the curriculum. Although doing so effectively may be possible in lower classes, it is not satisfactory at higher levels. Many countries have made attempts to retrain practising teachers so that they can become better teachers of mathematics and science in the upper primary classes. Such arrangements have rarely become institutionalized.

The rapid expansion of education in Africa has demanded that many teachers are trained over short periods. Inevitably this has led to a decline in quality. A survey of training programmes for secondary school teachers reveals wide variations between countries (Hanson & Crozier, 1974) and within countries over time (Fabiano, 1980, 1995), as illustrated in tables 9.6 and 9.7.

This applies to primary school teachers. Entry qualifications also vary, depending on demand — when large numbers of teachers are required either qualifications are lowered or training periods are reduced, or both. Unless the training experience is modified effectively, inevitably this leads to the production of mediocre teachers.

To address teacher quality, many countries have established, either temporarily or permanently, in-service programmes aimed at improving content knowledge as well as teaching methods (Mkaonja, Yadidi & Hau, 1994). Some have been instituted specifically to upgrade teachers' academic qualifications.

The success of the different approaches cited depends on the quality of teacher development programmes as well as the professional environment of schools in which the teachers will work. Because of the importance of the latter, all school staff, from principals to laboratory assistants, must be reoriented. Residential pre-service programmes are costly — especially so if there is a high attrition rate among qualified teachers. Some programmes are more costly than others such as, for example, degree programmes in Kenya in comparison with those of the Kenya Science Teachers' College and the Kenya Technical Teachers' College whose graduates are in high demand. Similarly, the delivery of in-service development programmes varies. Long residential programmes are often expensive, distance courses ineffective. School-based teacher development, such as those programmes being tried in Zanzibar and

South Africa, appear promising. Too often teacher development programmes involve 'topping up' teachers with content, and when this proves unsatisfactory, even more is provided.

However, for maximum cost-effectiveness, science educators must first diagnose classroom problems and then redesign teacher development programmes accordingly. Dyasi and Worth, Savage, Onwu and others discuss teacher development in more detail than is possible in this chapter.

**Table 9.6: Illustrative nondegree undergraduate programme, 1972**

| Country | Institution | Qualification | Entry point | Duration | Prac teach fieldwork (weeks) |
|---|---|---|---|---|---|
| Botswana | UBLS | Tchr Cert | CSC, Div II | 3 yrs* | 10 (100 %) |
| Ethiopia | Coll Tchr Ed<br>HSIU Dept Tech Ed | JSS Dip<br>Dip† | Postsec<br>Various, inc ESLC tests | 2 yrs EUS<br>2 yrs 2 EUS | 12 (15 %)<br>n/a |
| Ghana | STTC<br>ATTC | Spec Cert<br>Dip | GCE (O)<br>GCE (O) | 2 yrs<br>2 yrs | 12 (approx)<br>12 |
| Kenya | KSTC<br>Egerton | S 1 Dip<br>S 1 Dip (Ag) | 3 O level<br>3 O level | 3 yrs<br>3 yrs | 10–12 (11 %)<br>n/a |
| Lesotho | UBLS | Tchr Cert | CSC | 3 yrs | 10 (13 %) |
| Nigeria | ATTC (Ondo) | Nig Cert Ed (NCE) | 3 O level or Grade II (PT) | 3 yrs | 12 (10 %) |
| Swaziland | UBLS<br>UBLS | Tchr cert<br>Dip Ag Ed | CSC<br>CSC | 2 yrs<br>2 yrs | 10 (20 %)<br>10 (20 %) |
| Zambia | Kabwe ATC | Ed Dip (UNZA) | CSC or 4 O level | 2 yrs | 6 (15 %) |

Notes: * A two-year course at UBLS became a three-year course in 1974.
† Diploma issued in Industrial Arts, Home Economics.

**Source:** Hanson, 1974.

The preceeding paragraphs strongly argue that more effective learning is possible if, inter alia, schools are provided with appropriately trained teachers. This requires careful planning of training programmes which account for all variables that affect teacher supply and performance (Williams, 1979). Permanent in-service support structures should form a continuous feedback loop with pre-service training, identifying strengths to build on and weaknesses to eliminate in both components.

**Table 9.7: Output of physical science teachers who followed education programme from year I**

| Year | Dip Ed<br>3-year course | BEd<br>5-year course | Total |
|---|---|---|---|
| 67 | 6 | 0 | 6 |
| 69 | 5 | 2 | 7 |
| 71 | 17 | 4 | 21 |
| 72 | 16 | 5 | 21 |
| 73 | 26 | 3 | 29 |
| 74 | 22 | 0 | 22 |
| 75 | 17 | 7 | 24 |
| 76 | 3 | 1 | 4 |
| 77 | 7 | 1 | 8 |
| 78 | 0 | 2 | 2 |
| 79 | 3 | 0 | 3 |
| | Phased out | 5-year + 4-year course | |
| 91 | 0 | 21 | 21 |
| | | 4-year course | |
| 92 | 0 | 10 | 10 |
| 93 | 0 | 10 | 10 |
| 94 | 0 | 14 | 14 |
| 95 | 0 | 28 | 28 |

**Source:** Fabiano, 1980; 1995.

## Printed teaching and learning materials

Caillods et al (1995) associate provision of print materials with problems arising from design, printing and distribution. Curriculum materials vary in quality and relevance from the excellent to the obviously outdated and inadequate. Availability also varies from the widespread to the virtually unobtainable — indeed, Odhiambo (1993) has described the situation as bordering on famine. In some countries 'unofficial' materials, such as examination guides, are more popular with students (and often teachers) than official curriculum materials. The range between countries and over time in any given country can be narrow, from a single text, to comprehensive student books, worksheets, teachers' guides, enrichment materials and adequate libraries. In some countries provision is free, others levy a nominal charge and in some parents bear the full commercial cost. Solutions depend on initial conditions. Some issues that could be explored are presented in table 9.8.

**Table 9.8: Curriculum materials and possible responses**

| ***Quality and relevance*** | |
|---|---|
| High | Concentrates on supporting effective use of materials through in-service and school-oriented support. |
| ***Availability of basic texts*** | |
| High | Consider increasing the range of materials and selective recovery of some of the costs. |
| Low | Reduce costs, invest in effective distribution systems, subsidise purchase and delivery to under-served schools. |
| ***Use of alternative curriculum materials*** | |
| High | Analyse 'why' official materials are not preferred if they do not exist, invest in material development and improved assessment systems. Consider revised system of textbook production that is more sensitive to effective demand for text materials. |
| Low | Explore the options to extend the range of materials available. |
| ***Range of printed materials*** | |
| High | Invest in developing of enrichment materials, teacher's guides, and other language aids. |
| Low | Provide advice on coherent choice of core materials. |
| ***Cost per book*** | |
| High | Reduce costs to affordable levels, provide selective subsidies. |
| Low | Consider selective cost recovery. |

**Source:** Caillods, F, Gottelmann-Duret, G & Lewin, K, 1996.

The availability of locally produced books is a function of the health of local publishing capabilities. Where the local publishing industry is healthy, competition results in cheaper products. Some countries established parastatal publishing houses in association with curriculum centres to break the monopoly of multinational publishers. Ironically, doing so often led to government monopolies that have stifled the growth of local publishing industries. Donor schemes, such as the World Bank's support for supplying schools with textbooks, bring only short-term benefits to learners. Schemes such as that of CODE, Tanzania, where a national committee selects manuscripts, guaranteeing purchase for distribution in deprived communities, have led to an encouraging revival of local publishing capacities.

In countries where science teachers' associations have a strong leadership role, members write many books. The Science Teachers' Association of Nigeria (STAN) is a striking example. In addition to textbooks, active associations often produce other educational material such as newsletters, journals or bulletins. This complements textbooks by providing up-to-date information on content and teaching methods, both of which are important for effective teaching. Some associations, such as those in Malawi and Lesotho, produce newsletters for and by pupils that encourage positive attitudes towards science and technology.

Science clubs, competitions, school and community interactive museums and science fairs play a role in changing attitudes to learning and such activities deserve financial support. Mschindi, Shankerdass and others make a strong case for the support that media can make to science and technology education. *The Teacher*, a monthly supplement in one of South Africa's leading newspapers, has already demonstrated its effectiveness in aiding the professional development of teachers.

### Syllabuses, examinations and teaching approaches

There must be consistency between all elements of science and technology education. Too often, in Africa, goals call for problem-solving citizens and innovative teaching approaches, yet content selection and examinations present teachers with little choice but to cram their students full of facts. Examination and assessment systems have direct costs in setting and administration, and indirect costs in terms of teaching time forgone, and may represent a significant proportion of the overall costs of schooling. Multiple-choice paper-and-pencil examinations need not only assess the ability of students to memorize (Savage, 3). Experience in Kenya has shown that at primary and teacher training levels, items that test higher thinking skills encourage rather than discourage practical activity in schools. Research is needed to find out whether such items help to discriminate for selection more or less effectively than costly practical examinations that rarely contribute substantially to a variance in candidates' scores (Lewin, 1995b).

### Students as mobilizers of resources

Any consideration of resourcing education would be incomplete if it were to ignore what may be the most important resource of all, namely the learners (UNESCO, 1990). Conventional classrooms ignore students and treat them as passive recipients of knowledge. However, the situation changes dramatically when they become actively involved in their own learning. In such classrooms students become teachers, laboratory assistants, providers of materials from the immediate environment, advocates for local support through their parents, providers of positive role models for the increased participation of girls, and so on (Anamuah-Mensah, 1995).

## CONCLUSION

Increasingly authorities such as the Ministry of Health, Malawi (1991) and the World Bank (1996) are arguing that an educated population is more productive, and that this in turn leads to increased wealth. Thus, education is an investment for individuals, families and the nation (Hallak, 1990). Educational policies in African countries must maximize the already substantial investment made by all stakeholders in education to achieve increased economic productivity.

As Lewin (1987) suggests, there are in principle only three ways to ameliorate the problems of resourcing schools. These are through the allocation of an increased proportion of the budget (policy of expansion), greater efficiency in the use of existing allocations (through cost-saving reforms), and through the transfer of some costs from the public budget to individuals, the community, or the private sector (cost sharing). Which combination is possible depends on a range of factors. Since salaries comprise an overwhelming percentage of education budgets, and these salaries have become increasingly limited in their buying power, possible strategies are further limited. However strongly science and technology educators may believe in the efficacy of increased funding to education, it is unlikely that governments will allocate extra funds in the near future. A common response throughout Africa to increased access and falling standards is the growth of private education at all levels and an increasing number of students going overseas for their education.

Our challenge in public education is to use existing resources more effectively, and to develop innovative ways of increasing resources for all stakeholders. In doing so we should note that 'push models' of innovation often become unsustainable unless there is a complementary 'pull' from those identified as beneficiaries. Educators and researchers in Ghana, for example, have succeeded in attracting an industrial contribution of a small percentage of profits to a Science and Technology Fund. They did so by demonstrating the contributions university researchers and consultants can make to the industrial sector and by appealing to the professionalism of industrialists in curriculum development.

As ways of improving the effectiveness of existing resources to science and technology education, this chapter has proposed: (1) a redesign of courses at all levels to promote scientific problem solving using local community resources; (2) delaying specialization and thus saving on costly equipment; (3) a review of equipment to maximize its relevance and usefulness; (4) a development of school-based models of teacher education; and (5) more extensive use of the media to support science and technology education.

Finally, it is vital that those involved in planning make informed decisions on what percentages of financial resources are spent on each component, such as teacher expansion and development, print material design, production and distribution, laboratories and consumables, out-of-class activities, and so on. Too often such decisions are made as a result of public demand rather than after objective assessment of the situation.

## ACKNOWLEDGMENT

I wish to thank Jophus Anamuah-Mensah, Keith Lewin and Mike Savage for making significant contributions towards the improvement of this chapter.

## REFERENCES

Akyeampong, K & Anamuah-Mensah, J. 1993. The mole concept — Revisited at the tertiary level, *The Oguaa Educator*, 11(1), pp 11–28

Anamuah-Mensah, J. 1995. *The Race Against Under Development: A Mirage or Reality.* Monograph, University of Cape Coast, Ghana

Caillods, F, Gottelmann-Duret, G & Lewin, K. 1995. Science education provision at secondary level, planning and policy issues: Synthesis of an IIEP research project, Paper presented at the Policy Forum on Planning Science Education Provision at Secondary Level, Magaliesburg, South Africa

Caillods, F, Gottelmann-Duret, G & Lewin, K. 1996. Science Education Provision at Secondary Level in Developing Countries: Planning and Policy Issues: International Institute of Educational Planning, Paris

Fabiano, E. 1980. Science curriculum development for a developing country. Chapter 3 of an unpublished MSc thesis

Fabiano, E. 1993. Provision of science equipment and materials for the secondary school sector in Malawi. Feasibility Study Report, unpublished

Fabiano, E. 1995. Provision of science education in secondary schools in Malawi. Paper presented at the Policy Forum on Planning Science Education Provision at Secondary Level, Magaliesburg, South Africa

Hallak, J. 1990. *Investing in the Future: Setting Educational Priorities in the Developing World.* Paris: Pergamon Press, p 46

Hanson, JW & Crozier, DJS. 1974. Report on the supply of secondary level teachers in Africa: shifting the locus and focus to Africa, p 122, unpublished

Ivowi, UMO. 1995. Science education at the secondary level in Nigeria. Paper presented at the Policy Forum on Planning Science Education Provision at Secondary Level, Magaliesburg, South Africa

Jegede, O. 1995. The knowledge base for learning in science and technology education. Paper presented at the meeting on African Science and Technology Education: Towards the Future (ASTE '95), Durban, South Africa, 4–9 December

Lewin, KM. 1987. *Education in Austerity: Options for Planners.* Fundamentals of Educational Planning Series. Paris: IIEP, p 130

Lewin, KM. 1995a. Development policy and science education in South Africa. Reflections on post-Fordism and praxis. *Comparative Education*, 3(2), pp 203–22

Lewin, KM. 1995b. Comments on E Fabiano's paper on resourcing science and technology education. Paper presented at the meeting on ASTE '95

Makhurane, PM. 1995. The role of science and development in technology. Paper presented at the meeting on ASTE '95

Malawi Ministry of Health. 1991. Annual Health Statistics

Mjojo, CC. 1994. Investing in research and development in science, engineering and technology for accelerated development. Paper presented at the Round Table on Science and Technology Protocol of the African Economic Community, Mangochi, Malawi

Mkaonja, M, Yadidi, DC & Hau, SA. 1994. Report on a study tour to Botswana and Swaziland, unpublished

Odhiambo, TR. 1993. Scientific and technological literacy for all: A perspective reality. *International Forum on Scientific and Technological Literacy for All:* Final Report. Paris: UNESCO

Onwu, G. 1995. Teaching large classes. Paper presented at the meeting on ASTE '95

Passi, FO. 1990. Planning for the supply and demand of qualified teachers in Uganda. *International Review of Education*, 36(4), pp 441–52

Ross, AR & Lewin, KM. 1992. *Science Kits in Developing Countries: An Appraisal of Potential.* Paris: IIEP

Savage, M. 1995. Curriculum innovations and their impact on teaching of science and technology. Paper presented at the meeting on ASTE '95

UNDP. 1994. *Human Development Reports, 1994.* Oxford University Press, New York

UNECA. 1994. Report of the Round Table on the Science and Technology Protocol of the African Economic Community, Mangochi, Malawi

UNESCO. 1990. World Conference on Education for All: World Declaration on Education for All and Framework for Action to Meet Basic Learning Needs

UNESCO. 1994a. A statistical profile of education in sub-Saharan Africa in the 1980s. Paris: UNESCO

UNESCO. 1994b. *Donors to African Education*. Paris: UNESCO

Williams, P. 1979. *Planning teacher demand and supply*. Paris: IIEP, p 51

World Bank. 1996. *The Economist,* January 5, pp 107–9

Yoloye, EA. 1989. A national survey on resources for science. *Mathematics and Technical Education in Nigerian Secondary Schools,* p 111

 

# 10

## The knowledge base for learning in science and technology education

*Olugbemiro Jegede, University of South Queensland, Toowoomba, Queensland, Australia*

### ABSTRACT

An appropriate and efficacious knowledge base is paramount for science and technology learning in Africa. This chapter examines types of knowledge and ways of knowing; local cultural and indigenous knowledge systems versus the universality of Western science; second and third-language teaching of students whose mother tongue is not English; teaching classes with students of many mother tongues; cognitive styles, constructivism, and concept learning in the African child; the African child's background; the impact on learning of belonging to rural versus urban communities, and the particular cognitive problems facing girls.

### INTRODUCTION

> If Africa is to make progress in moving from the eighteenth century into the late twentieth century, unconventional approaches to science and education unprecedented in world history will have to be devised. (Fafunwa, 1967)

Professor Babatunde Fafunwa's statement was revolutionary at a time when many African countries were colonies or had just gained their political independence. Almost three decades later, not much has changed. In the twilight of the 20th century, Africa has made little progress in teaching science or technology, neither devising anything unprecedented nor evolving any unconventional approaches. Indeed, the continent is still groping on its educational, scientific and technological journey into the 21st century. While world history has seen unprecedented achievements in science and technology, made largely in the West, African people show little concern about the less-than-acceptable performance of their continent. Furthermore, neither

Africa as a whole nor individual African nations have charted a markedly different course for future development.

There are reasons for the unenviable state of science and technology education in Africa. Colonial educational tenets may have been poorly assimilated and are still seen as foreign. Badly prepared teachers may contribute to poor student achievement. With an illiteracy rate of about 75 %, society may lack an understanding of what school science means to the individual, the local community or the nation. Should we expect underresourced, overcrowded classrooms in dilapidated school environments to produce the scientific and technological geniuses for whom we yearn? Should Africa's goals for teaching science and technology be different from those in the Western world? Should Africa devise its own relevant and culturally responsive approaches rather than adopt science and technology curricula from other parts of the world? In a continent where many governments experience economic crises, are ridden with fraud, and at best pay lip-service to education, is it realistic to expect world-class achievements in science and technology? What should we expect from a continent whose higher institutions are in decline, with outdated, understocked libraries, weak undergraduate programmes, and uninterested and uncared-for post-graduate students? What can we hope for from countries abandoned by many of their best academics for laboratories and universities in the West? How realistic are our expectations from an investment in an area that is poorly understood even by those who 'own the knowledge'? What can we expect when the culture of Western science taught to our children contradicts their indigenous culture and world-view? Is it any wonder that Africa is yet to produce revolutionary discoveries in science and technology that will rival those of the West?

It may be unfair to expect more from a continent where effective contact with Western science and technology is less than a century old. Yet Africa has participated as an equal partner in global movements in science and technology education. The innovations of the 1960s and 1970s put Africa on the educational world map, thanks to science educators such as Fafunwa, Dyasi and Yoloye. The achievements of scientists working in Africa, such as Odhiambo and Onabamiro, are widely recognized, and the work of African scientists, technologists and educators in prestigious institutions in the developed world is evidence that Africans can equal those who brought Western science and technology to the continent.

Perhaps Fafunwa's vision was premature. The pace of the journey towards achievements that would be 'unprecedented in world history' has slowed. Conflicts, disasters (both natural and self-inflicted), a lack of positive social transformation, unstable and despotic leadership, an absence of comprehensive development policies or a failure in their implementation, a lack of political will, and a general level of poverty — these are some of the problems that have slowed down the journey towards scientific and technological development in Africa.

We need to consider what the African Academy of Sciences called the development of a science culture in Africa (Tindimubona, 1991). This should include a resolution of issues such as indigenous knowledge systems and traditional education; the knowledge base for learning; public understanding of science and technology; a definition of science and technology for Africa; how these subjects should be taught, and how to popularize science and technology for different target groups in all parts of Africa.

This chapter considers the appropriate knowledge base for science and technology education in Africa. It goes on to discuss the structure of knowledge, ways of knowing, the cultural context of science and technology learning, and, finally, it suggests what we might do to improve the situation.

## THE TRADITIONAL KNOWLEDGE BASE

Both the structure of a discipline and its body of accumulated knowledge play an important role in learning. 'Knowledge base' is a term used differently by different disciplines. Cognitive science, expert systems, and artificial intelligence studies frequently use the term. In psychology and education the term is used to depict the 'distillation of understandings from experts, narrative views, and meta-analyses of variables that influence learning' (Wang, Haertel & Walberg, 1993: 253). Varying contexts and social situations nurture a knowledge base that accumulates over time. This chapter takes the view that a knowledge base it is not only a distillation of ideas as defined by Wang and associates; rather, it is an accumulation of information and practices from which learners can draw to aid further learning. It is therefore content-oriented and affected by context. A knowledge base should encompass information derived from the instructional, sociological, anthropological and psychological elements of a society. In Africa, the knowledge base for schooling should draw from traditional and current beliefs, taboos, superstitions, customs and traditions. From the Western view, a knowledge base includes only evidence that can be transformed empirically into knowledge (Hedges & Waddington, 1993; Kerderman & Phillips, 1993) and that experts deem credible. This excludes the learner's context. To teach science and technology in African schools within such a narrow definition is to ignore what catalyses learning within the student's environment.

According to Gagne (1975), knowledge acquisition, the individual's construction of reality, and the ability to think are all dependent on growth, learning, and their interaction. Contemporary theory looks at learning and memory as information processing, gives consideration to thinking processes, links knowledge and performance, and attempts to explain problem solving. Cognitive research has shown that: (1) context is important to understanding; (2) learning is not automatically transferred to new settings; (3) passive learning is not conducive to developing cognitive and metacognitive skills; and (4) higher-order learning is not a change in behaviour but the construction of meaning from experiences (Thomas, 1992). As elaborated by Resnick (1989), learning is a process of knowledge construction, not of knowledge recording or absorption. Learning is knowledge-dependent, and the learner uses existing knowledge to construct new knowledge. Whereas modern cognitive psychology views learning as a process that results in knowledge being stored in compartments of the mental schema, in Africa learning is viewed as a holistic process governed by a knowledge base that includes both factual knowledge and beliefs and customs.

 

The knowledge base for science and technology consists of the conceptual, skill, social, and resource domains. The conceptual domain is built by using pupils' background experiences, devising relevant examples and linking learning with the dominant cultural world-view. The skill domain is developed by creating opportunities for learners to use the process skills of science. The social domain is developed by involving learners as active members of the scientific community through group work and communication using appropriate reporting. The social domain is built by exposing students to real problems and encouraging innovative responses. Finally, the resource domain is built through access to appropriate materials for exploration and interpretation.

## CONTEXTUAL LEARNING AND CULTURE

As the construction of new knowledge, learning is dependent on the existing knowledge base. Both old and new knowledge are contextual, as defined by Brown, Collins and Duguid (1989), Connelly and Clandinin (1990), and Martin and Bouwer (1991), among others, who stress the situated nature of cognition. Within the African context, as in any other context, situated cognition cannot be separated from the sociocultural environment.

The sociocultural factors of a learner's environment significantly affect achievement in school work (Biesheuval, 1972; Jegede & Okebukola, 1988, 1989; Jegede, 1995a and b). Glaser (1991) asserts that cognitive activity is inseparable from its cultural milieu. This has been supported by anthropologists such as Ogbu (1992), who found that school learning and performance are influenced by complex social, economic, historical, and cultural factors.

Every society educates the younger generation as a means of passing down its sociocultural attributes. These attributes largely control what a child learns and becomes (Ogunniyi, 1988a). Culture subsumes all we undertake: even science and technology education is a human enterprise that involves the transmission of cultural heritage (Gallagher & Dawson, 1988). Cossons (1993) argues that since science is a human activity and a central element of culture, when we try to understand how people learn science and how scientific knowledge is structured, we should first understand its cultural context. In support of the need for cultural studies in science education, Cobern (1993: 55) suggested that educators should understand the 'fundamental, culturally based beliefs about the world that students bring to class, and how these beliefs are supported by students' cultures, because science education is successful only to the extent that science can find a niche in the cognitive and sociocultural milieu of students'. Technology educators must also recognize the role of indigenous knowledge in teaching and learning (Swift, 1992).

Two major trends in science and technology education will significantly affect teaching, learning and research in the coming decades. The first is a shift from notions of science as the rigid, 17th-century, positivist, 'Royal Society' view to notions of science as a cultural enterprise practised by all human beings within a social environment. The second is the recognition that pupils bring alternative frameworks from different cultures into their science learning. This view supports the need for using indigenous technologies in teaching (Swift, 1992; Loving, 1995).

The realization that culture plays a central role in science and technology education, especially in an environment where Western science is seen as a foreign culture, has prompted a proliferation of studies (see, for example, Jegede, 1989, 1994, 1995a, 1995b; Jegede & Okebukola, 1989, 1990, 1991, 1992, 1993; Jegede & Olajide, 1995; Jegede, Fraser & Okebukola, 1994; Okebukola & Jegede, 1990; Ogawa, 1986, 1995a; Ogunniyi, 1987, 1988a, 1988b; Ogunniyi, Jegede, Ogawa & Yandilla, 1995). Scholars have looked at a number of issues including factors that affect science learning in non-Western cultures, cosmology and science learning, science as a foreign culture, the influence of traditional culture in science classrooms, and how to measure the sociocultural environment in science classrooms. A notable outcome of some of these studies has been the identification of authoritarianism, goal structure, traditional world-view, societal expectations, and the sacredness of science as predictors of sociocultural influence on learning and teaching science. This type of research is currently gathering momentum: perhaps educators in other non-Western countries will recognize its significance as an approach to understanding science and technology learning.

## World-view and duality of cultures

The world-view on which science and technology education is based has two main aspects (Cobern, 1993). The conceptual aspect concerns how individuals in a particular environment perceive knowledge. The social aspect concerns how individuals negotiate knowledge in their society. These aspects, or 'ecologies', have been referred to as 'eco-cultures' (Okebukola & Jegede, 1990) or 'conceptual eco-cultures' (Jegede, 1995a), and have been the focus of many studies on sociocultural factors. Cobern, who has been instrumental in the study of world-view in science education, defines world-view as the 'culturally dependent, generally subconscious, fundamental organization of the mind that manifests itself as a set of presuppositions that predispose one to feel, think and act in predictable patterns' (1993: 58). His definition implies that world-view precedes and forms the cognitive background for both modern science and indigenous knowledge. It also implies that Western and non-Western conceptual systems are grounded in different world-views. Aikenhead (1996) reminds us that science itself is a subculture of Western culture. School science and technology as currently taught in Africa are based on one type of world-view — the Western world-view — that claims to be superior to others.

I have used the term 'Western science' to represent the science taught in schools throughout Africa. The term 'Western' identifies the science that dominates the world, has become the basis for technology, and is often labelled as 'modern'. A widespread misconception is that 'modern' is synonymous with Western and superior. 'Modern' is often used, especially in Western cultures, in opposition to 'traditional'. Since most non-Western societies are traditional, they are therefore considered non-modern and dependent on Western culture. The terms 'modern' and 'traditional' are frequently taken as opposites in Western cultures, but this is not necessarily the case. That a culture is traditional does not mean that it is not modern; that a culture is non-Western does not make it dependent or inferior.

Those who argue that there is no such thing as Western science tend to claim that the scientific culture that evolved in the West is universal and must be imposed on other, 'traditional' cultures. Such a process may cause as much damage to African scientists, however, as colonialism wrought to the psyche of the colonized. The insensitive imposition of this so-called universal science in a manner that is indifferent to the indigenous knowledge base implies acceptance of an alien protocol in understanding what some call reality (objective or otherwise). My own use of the term 'Western science' derives from a sociocultural theory that I have labelled the 'ecocultural paradigm'.

There are many cultural differences in how science is perceived and learned. The science that students are taught in African schools is not indigenous to them, but rather is imposed from outside. Colonized non-Western countries have no choice but to adopt, as if it were their own, the science that comes with Western culture. The Western view obliterates their indigenous ways of knowing: many Africans educated in science within a Western framework find it difficult to shed the baggage imposed by such imperialism.

Western science is one tool the human mind can use to explain the physical world, but not the only one. However, in my opinion, through imperialism, coercion and persuasion, Western science has come to be seen as universal science.

If we accept that science is a human attempt to understand nature, then every culture has its science and scientists. We teach the Africanized view of Western science in African schools. The learner in African classrooms is therefore faced with two cultures, each arising from different world-views: the culture of science and the culture of the local environment. A third dimension is that, through the colonization process, the Western science culture brought to Africa and transmitted through the culture of the Western world (Aikenhead, 1996) demands that the learner also acquires the culture of the West. In effect, an African learning science has to cope with two world-views and three cultures! It is not surprising that few outstanding African scientists, technologists, and science and technology learners have emerged. Many non-Africans claim that the distinct cultures of science and society affect people in the Western world as strongly they do people from non-Western societies and that Africans focus too strongly on the issue. I contend that the African situation is different, that Africa has a single world-view, and that generalizations about Africa are justifiable.

Africa has its science and technology that are not taught in schools. School science and technology are taught in African classrooms as a subculture of Western cultures (Pomeroy, 1994; Phelan, Davidson & Cao, 1991). Aikenhead (1996) claims that, because science is a subculture of Western culture, it is not as foreign to Western learners as to Africans who learn science and Western culture while living within an indigenous world-view. The difference between the Western and the African learner of science and technology is thus a matter of kind and of intensity. The differences that the learner in the West experiences in the science classroom are similar to those experienced by society. For African learners in science classrooms, the differences take them from their indigenous cultures. While this may appear to be categorizing science as school knowledge (Cobern, 1995a), it is more than that. Africans not only categorize science as school knowledge, but also as the culture that school science represents and which is foreign to their world-view. If world-view is an antecedent to cognition, then communal organization, theory of knowledge, concept of causality, authoritarianism, goal structure, kinship system, story telling and riddles, and worshipping of ancestral spirits as part of the African world-view must significantly impact on how both science as school knowledge and science as Western culture are viewed by the African learner. Wiredu (1980) agrees that Western science and technology have alienated Africans from their culture. He describes the phenomenon of 'belonging at once to two worlds, . . . a new dualism . . . that causes a kind of ethnic schizophrenia in some spheres of conduct' (1980: 23). The African is operating in both these worlds as best as he or she can. Any individual faced with a similar problem anywhere would possibly respond in the same way as the African, especially if that individual were not adequately prepared to cope with the conflicting realities of life' (1980: 7). Abimbola (1977: 23) writes: 'The problem is that the African child comes to the school with a load of mysteries that plague his mind. If care is not taken these mysteries, usually tagged as "superstitions", are capable of causing blockage to any scientific knowledge the child might acquire as a result of schooling. So, even when the child has a reason to believe the scientific explanations of a particular phenomenon, his deep-rooted African world-view may lead him to regard the explanations as a bundle of neatly fabricated lies.' The duality of views with which African learners grapple must be effectively resolved if science and technology are to progress. Attention needs to focus on what happens when cultural traditions clash with science and technology in African classrooms. Even in Western environments where the debate about multicultural education has emerged this is a relevant question.

A related and often contentious issue is the question, 'Is there an African world-view?' Non-Africans, and indeed some Africans, wonder if Africans share a unified culture. Africa has 54 countries, over 650 million inhabitants with over 500 languages and ethnic groups. How can one therefore say that Africans share a common world-view?

The original inhabitants of Africa were hunters and gatherers who moved from one part of the continent to another. Population growth resulted in kingdoms, chiefdoms and, with the arrival of the Arabs, emirates. They shared certain characteristics due to their common experiences of precolonial trade in goods, crops and slaves, as well as a common ancestry. What now constitute the 54 countries of Africa are artificial boundaries dividing cultures and families. They were created by colonial powers in the 16th and 17th centuries and formally ratified at the infamous 1884 Berlin Conference without the consent of the people.

I subscribe to a pan-Africanist view about the unity of African culture, but add that I believe that there are differences at the micro level. Forde (1954) observed that the material and cultural backgrounds of the indigenous peoples of Africa have led to common beliefs and attitudes. Idowu (1963: 103), in an attempt to answer critics of pan-Africanist world-view homogeneity, says that observation and comparative discussion with Africans from various parts of the continent 'will show, first and foremost, that there is a common factor ... and common Africanness about the total culture and religious beliefs and practices'. Mbiti (1969) established that concepts of witchcraft and traditional medicine are shared by all African societies. Abimbola (1977) concluded that, in spite of minor differences in the ways African communities look at nature, there are similarities that can justify speaking of an African world-view. At the macro level most African communities have similar beliefs, customs and traditions relating to theories of knowledge, causality, religion, concepts of time and space, kinship system, rituals, marriage celebrations, witchcraft, ancestor worship, reincarnation, story telling, and so on. These constitute an African world-view that is shared by most cultures of sub-Saharan Africa. Differences are of degree rather than kind. One example is the African naming ritual. The celebration includes festivities and ancestor worship, involves the whole community, and names have special meanings. There may be differences as to whether the baby is named a week, a month or three months after birth, or whether the names relate to the mother's or the father's family. A second example is that in most African communities marriage is a communal activity involving whole communities or villages. Most African communities practise some form of dowry payment. Differences concern whether the groom's or the bride's family pays the dowry and the form of payment.

The diverse African world-views share four fundamental features: (1) a belief in the existence of the Creator — the supreme God; (2) a belief in the continuation of life after death — reincarnation; (3) the human being as the centre of the universe; and (4) a theory of causality. These constitute an anthropomorphic view of nature that governs how Africans think, the way they act, the way they relate to one another, and are the sociocultural antecedents of how Africans learn science and technology. According to Glaser (1991: 132), 'the way students represent the information given in a mathematics or science problem, or in a text they read, depends upon the structure of their existing knowledge. These structures enable them to build a representation or mental model that guides problem solution and further learning'. African learners use an African rather than a Western world-view to build enabling structures to understand nature and school science. Using the logico-structural model of world-view categorization borrowed from anthropology by Cobern (1993), it is possible to differentiate between African and Western world-views, as set out in table 10.1 opposite.

**Table 10.1: Comparison of African and Western world-views using the logicostructural model**

| World-view categories | Sub-category | African world-view | Western world-view |
|---|---|---|---|
| **Non-self** | the super-natural | common religious beliefs | privatized religion |
| | the natural | anthropomorphic; monistic/vitalistic | mechanistic; empirical/ theoretical |
| | the social | sage practice; oral culture; communal learning | 'questions authority', written culture; individual learning |
| **Self** | group | strong social cohesion | weak social cohesion |
| | individual | communal good takes priority; individual is a contributor to communal goals | realization of personal goals given priority |
| **Classification** | knowledge acquisition | determined more by age, and community structured | realization of personal goals given priority |
| | materials | derived from nature for all circumstances | different classificatory systems |
| **Relationship** | | communal: goal structure; deference to sacred sites | individualist and competitive; nothing is sacred |
| **Causality** | role and place of person | victim regarded as constant; every event ascribed a cause; elements not relevant to each other | victim and circum-stances regarded as variables in a hypothetico-deductive fashion |
| | accidental occurrence | can be observed | conjunctions with no laws |
| **Time** | solutions | appeasement/purifi-cation of the system | through education |
| | life | cyclical, continuous flow; present in everything; reincarnation ensures relationship | linear; time is in separate units and looks towards the future |
| **Space** | physical | everything, including those invisible, is one reality | must be visible to be real |
| | spiritual | intangible but very important; everything has a god | not considered an objective assessment of reality |

## CONSTRUCTIVISM AND SCIENCE AND TECHNOLOGY IN AFRICA

Several interrelated points emerge from this discussion of world-view, culture, and science and technology learning. First, meaning is affected by the viewpoint of a culture. Second, social interactions within the community define meaning. Third, although meanings are socially determined, the individual uses an idiosyncratic pattern to construct meaning. Thus, when engaging in social interaction while attempting to make meaning personal, an individual experiences an interplay between cognition and affect through a world-view that serves as an interpretive framework. The learner's understanding of any new meaning is strongly influenced and determined by prior knowledge that is in turn determined by cultural beliefs, traditions and customs governed by a world-view. If prior knowledge exists as a result of cultural beliefs and theories, then different groups are likely to have different prior knowledge (Driver & Erickson, 1983; and Snively, 1989). This will affect the way a learner creates meaning as well as the way that the different cultures of science and technology, including that of Western science, are viewed by an African learner.

Constructivism — which underlies much of current thinking about science education — emerged from a convergence of three major areas of research (Solomon, 1994). These are the theory of personal constructs (Kelly, 1955), the notion of 'Children's Science' (Driver & Easley, 1978; Von Glasersfeld, 1989; Osborne, Bell & Gilbert, 1983), and the social construction of knowledge (Vygotsky, 1978; Wheatley, 1991; Cobb, 1989; Solomon, 1989). Conceptual change research has dominated constructivism and deals with the key role of students' prior knowledge (Solomon, 1989) and the reflective process of interpersonal negotiation of meaning.

Cultural anthropologists and social constructivists have proposed a theory that knowledge is socially negotiated and that a learner's background and prior knowledge influences school achievement (Prawat, 1993). This call for recognition of a learner's sociocultural background in teaching science and technology has gathered support from many sources (see Driver, 1979; Cobern, 1994; Atwater, 1994; Jegede, 1995a; Ogunniyi, 1988b; Solomon, 1989), and draws on the work of Piaget (1970) and Vygotsky (1978) which pointed out that all learning takes place in a social context. The social context acts as scaffolding, providing assistance that fosters co-construction of knowledge while the learner interacts with other members of society. Wertsch and Toma (1992) suggest that a sociocultural approach to mediated learning should be adopted. This approach claims that mental functioning is inherently situated in cultural, historical and institutional contexts. In Africa, day-to-day interactions and explanations of natural occurrences are influenced within the sociocultural environment by philosophical and religious beliefs, a theory of causality, taboos and superstitions. These impact on the attitudes, thoughts and behaviours of pupils as they learn and understand science and technology and apply their learning.

People everywhere, including Africa, socially negotiated ideas about reality before Jean Baptiste Vico philosophized about constructivism in 1610, or Von Glasersfeld extended Vico's idea to include radical constructivism. African pupils construct their understanding of nature on a daily basis using their world-view as prior knowledge.

Children develop meaning on the basis of their interaction with elders, nature, and views of their peers. African children learn about their environment using prior knowledge situated within their non-Western world-view. Problems arise when they are asked to learn Western science and technology in school together with Western culture. A recent on-line discussion contribution by Ken Tobin and forwarded by Alejandro Gallard (see e-mail posting to RESODLAA, Saturday 23 September 1995) has confirmed my thoughts about how learning fails when students enter the classroom. One reason may be that prior knowledge presents itself to students either as capital or as a handicap. The knowledge-as-capital metaphor allows learners to examine the viability of the ideas presented and to construct meaning without major hindrance. Prior knowledge situated within the African world-view becomes a handicap when a Western world-view is used as a framework for learning science and technology. The learner experiences mental perturbations and cognition is impeded. What we therefore regard as learning by the African child is an accumulation of information compartmentalized in mental schema to be used during examinations or when issues of indigenous knowledge are raised. Most problems arise when the ethos, values and mores of the two communities clash in science and technology classrooms. Neither world-view is presented in ways students can understand, participate in or use for the construction of knowledge. Tobin (1995: 1) says they cannot 'co-participate in a shared discourse and their discursive resources are frequently considered as having little or no value either way'. Driver, Asoko, Leach, Mortimer and Scott (1994: 11) succinctly assert that 'learning science in the classroom involves children entering a new community of discourse, a new culture'. It is like entering a conversation mid-stream and expecting to contribute to it when you neither know the rules nor are well informed about the issue being discussed. In the case of African learners, the prior knowledge they bring into such a discussion is not in consonance with the philosophy, orientation and knowledge base of the issue being discussed.

## LANGUAGE AND LOCATION ISSUES

Language plays an important role in teaching and learning science and technology. In most schools in sub-Saharan Africa, English, French or Portuguese is the official language of instruction. These languages have been adopted as convenient alternatives to the presumed controversy that might arise in the choice of one of the native languages. Those who support the use of a foreign language argue that it is universal, economical and has been tested and found viable. Those who oppose foreign tongues claim that school subjects can be taught using indigenous languages, that using a foreign language is elitist and that its use alienates children from their culture.

Linguistic barriers are cited as impediments to successful acquisition of science and mathematics knowledge by students from non-Western cultures (Hodson, 1992). Rampal (1994: 132), commenting on the situation in India, says that 'many students never make it to high school because an emphasis on rote memorization of remote concepts in a formidable foreign language alienates the majority of young children, and they drop out long before they complete elementary school'. This is typical of what occurs in Africa. Both Bamgbose (1984) and Collison (1974) have shown that scientific concepts are best learned and understood in the students' mother tongue in spite of its technical limitations. My experience is that pupils often have to translate mentally the science and mathematics learned in English into the mother tongue for meaningful understanding to result. Prophet (1990) has stated that language is part of a larger, more complex issue. He asserts that 'language is not merely an incidental way of communicating, or solving problems, or reflecting. Rather, in our rational reconstruction of reality, language acts as the mediator and supporter in the continuous matching and fitting that takes place between "things as they are" and "things as we know them"' (1990: 21). The issue of language complicates learning and teaching science and technology in Africa where the subject, the culture of the subject, the language of instructing the subject, and the language of discourse are all unfamiliar to the learner.

Many studies show that African children in urban schools significantly outperform their rural counterparts on achievement outcomes (Jegede, 1995a; Jegede, Naidoo & Okebukola, 1996), though differences in their world-views are insignificant (Jegede & Okebukola, 1989, 1990; Okebukola & Jegede, 1990). It appears that the urban environment alone supports achievement (as opposed to learning) in science and technology. This needs further investigation.

## GENDER AND SCIENCE AND TECHNOLOGY IN AFRICA

Gender inequality in science and technology is a worldwide phenomenon. Little is known about the factors that influence girls to choose or reject science and technology (Catsambis, 1995; Baker & Leary, 1995). Many intervention programmes have been implemented to address the differential achievement and enrolment in science and technology to: (1) demasculinize and demystify science; (2) implement teaching strategies that actively involve students; and (3) improve girls' confidence and self-perceptions of their ability to tackle science and technology. Unfortunately, as Kahle and Meece (1994) have reported, the gap between male and female performance and interest in science appears to be on the increase in spite of these efforts. If the issue constitutes a serious problem in developed Western societies, it is worse in Africa. For example, with 100 million inhabitants, Nigeria has the largest population in Africa. Though about 60 % are female, the Science Teachers' Association of Nigeria (STAN, 1992) reported that less than 30 % of the one million girls in secondary schools take science, only 6 % of those who enrolled in the West African and the senior secondary school certificate examinations are girls, and that women constitute less than 10 % of the total enrolment in Nigerian universities for science- and technology-based disciplines and less than 5 % of the science faculty in Nigerian universities.

Gender inequity in science and technology is pronounced in Africa where socio-cultural factors contribute to achievement and attitude differences. To date little has been done to narrow the gap. Three separate studies (Jegede & Okebukola, 1989, 1992; Okebukola & Jegede, 1990) in Nigeria — a country that is predominantly traditional — revealed similarities between males and females in their perception of four out of the five social-cultural factors investigated in science classrooms. In a study comparing the preferred and perceived science classroom environment, students, irrespective of gender, perceived and preferred the sociocultural environment of their classrooms in a similar way (Jegede, Agholor & Okebukola, 1996). These results corroborated those of Cobern (1995a), who found that neither gender nor achievement in science is correlated with the concepts ninth graders typically use in discussion about the natural world. These results indicate that world-view usage in science and technology classes is not gender but environment dependent.

Women and girls think that African societies have a low regard for their ability to perform in science and technology (Jegede & Okebukola, 1992). This affects girls' motivation to choose science-based careers and supports the widespread view of the domestic role of women in traditional African society. Women perform tasks that include household chores, child rearing, feeding the family, and educating infants. Men spend their time on the farm or working for money and claim to have little time or energy for household matters. Masculinity is revered and the male macho image rules. The roles defined for women are subservient and their menial jobs negatively affect their self-image. Women are to be seen and not heard in most African societies and are deemed secondary to males in many cultural matters. Women's roles are trivialized and there are limited expectations and recognition of their contributions to development, the knowledge base, and of possible careers and occupations. In classroom situations, the same views affect girls' achievement in, and attitude to, science and technology studies. Consequently, an achievement gulf continues to exist between males and females in formal school settings. However, the debate continues in Africa as to whether education should change perceptions of the role of gender and whether gender issues should be pursued with the vigour that feminism is in the West.

## FACING THE PROBLEMS SQUARELY

This paper has analysed why science and technology have not brought the expected changes in Africa in spite of expectations and the commitment of resources to their teaching. Barriers identified have included: (1) the traditional African knowledge structure that is widely believed to be nonlinear and multifaceted rather than hierarchical and pyramidal, though empirical information is lacking; (2) the lack of an appropriate knowledge base derived from the African world-view; (3) that African learners in science and technology classrooms must deal with a duality of world-views and a multiplicity of cultures; (4) prior knowledge from the indigenous African culture that acts as a handicap in the construction of school knowledge in science and technology; (5) language differences that also act as an impediment to learning; and (6) insufficient attention being paid to issues such as gender in science and technology learning in Africa. Having identified the problems, we should consider what Africa can do to ignite, in the spirit of Professor Fafunwa, an unprecedented revolution in education. How can Africa make its science and technology education more practical? How can we sow the seeds that will bring Africa to international achievements in science and technology? Africa already has its path charted. It is part of a developing world culture and the idea of reverting to the precolonial era must be discarded. While becoming a part of the new world, Africa can and should chart its own path to development through science and technology. The answer and the motivation lie within Africa and with African educators in partnership with committed external organizations and concerned friends of the continent. We should examine the reasons for hoping that the future will be kinder to Africa than the past.

## WHICH SCIENCE, WHICH WORLD?

In constructing an appropriate knowledge base for science and technology in Africa, we must resolve the conflicting views held about science and indigenous knowledge. Science and technology have begun to accommodate varying views regarding their utility and place as social institutions. However, the continuing dominance of positivist notions of science guided by a strict adherence to empiricism, and of Popperian ideas regarding the methods of science, require adjustment to fit current philosophies about knowledge and learning. For some time those who determined science and acted as gatekeepers to the community of scientists assumed an unnecessarily divisive posture. People everywhere are intimidated by science, with serious consequences. But since the days of Kuhn and Feyerabend, monolithic thinking about science, its rigidity and its unified structure have given way to other paradigms. Science is now seen as an evolving, disciplinary matrix (Loving, 1995), as an evolving way of coping with the world within specific contexts and cultures. A consideration of contexts and cultures means one must look at alternative world-views. However, mainstream Western science remains defensive when issues of other world-views, or considerations of science within indigenous cultures, are raised. Indigenous knowledge is frequently written off as myth, superstition and folklore and not regarded as science by a group who would not like to see change.

**Figure 10.1: Ways of treating science and world-views**

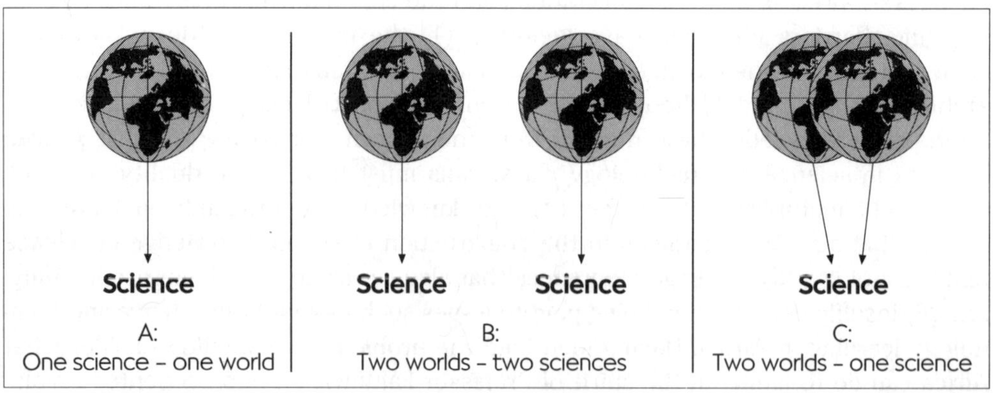

Africa must decide whether: (1) school science and technology should adopt Western science within a single world-view (figure 10.1A); (2) to allow, as is currently the case, the compartmentalization of two world-views in which school science exists side by side with the indigenous knowledge of learners (figure 10.1B); or (3) to acknowledge the existence of two world-views and to try to integrate them so that a common explanation of science and technology concepts becomes possible (figure 10.1C).

Science and technology should be seen as ways to understand the natural environment and their study must be integrated with the world outside the classroom. Some scientists and technologists believe that science should be practised for its own sake and deny responsibility for how its results are applied. Such beliefs are increasingly challenged by society.

Recently, Cobern (1995b) looked at the issue of the separation or integration of science with world-views and invited my comments about world-view language games. My response (Jegede, 1995c) was emphatic support for the view that to make learning meaningful, there must be integration of science knowledge with the learner's world-view. If science is observing and understanding everyday life, science and everyday thinking should not be radically different. Separation occurs because of limited and elitist definitions, and the hegemonic hierarchy of the knowledge structure of science. Science became institutionalized by the Royal Society in 1662 and the elitist problems created still exist in schools. Our attitudes and perceptions of science are coloured by a range of factors that include the rigidity with which scientific knowledge is dichotomized as abstract–concrete to give an impression of ports of entrance into its court. Science is seen as monolithic because of the way scientists believe that their methods are beyond questioning. But there are several ways to get to the market as people in my culture say. Getting to the market is more important than the road one travels, if other variables are of no consequence, and science too should not be seen as a single path but as an evolving map of ways to cope with the world.

As the chameleon's skin changes colour, so, within the world-view language game, a learner can do many things with any concept. The chameleon survives by responding to its environment. In learning science and technology the response of non-Western learners is to blend their world-views to enrich understanding. At first they wrestle with the two world-views, then they compartmentalize the Western idea in their cognitive system, to be used on appropriate occasions in the science classroom. Cobern (1994) calls this cognitive apartheid. I propose a theory of collateral learning to explain the degrees and hierarchies of 'cognitive apartheid' (Jegede, 1995a). Aikenhead (1996) labels this stage of cognitive apartheid as 'assimilation', indicating that the subculture of science is at odds with the world of students' life-world cultures. As soon as non-Western learners of Western science leave school, they shuffle the cards and, with luck, bring out the one required to make sense in traditional society. At the highest level, when operating within traditional society and confronted with two opposing world-views, we question the wisdom of compartmentalization and see the world as a unity. My theory of collateral learning attempts to explain this. The duality in the mental schema of non-Western learners with a resilient indigenous knowledge framework when they learn Western science in school results in collateral learning. Collateral learning represents the process by which a non-Western learner in school constructs, side by side and with minimal interference and interaction, Western and traditional meanings. Collateral knowledge is therefore the declarative knowledge of a concept that such a learner stores in the long-term memory, with a capability for strategic use in either the Western or the traditional environment (Jegede, 1995a).

The world-view language game cannot be played without tension, because the games are played on the same field. The tension leads to a mental reorientation of the person forced to 'assimilate' and use these views. Non-Western learners must shuffle the language cards each time to play the game well. They may say '. . . if I'm in a science classroom then it has to be this card, if I'm in the palm wine bar it must be that card', and so on, just as the chameleon changes skin colour for its survival. Mostly this works, but doing so stifles initiatives to understand or predict the environment. Does the chameleon have a knowledge base to help cope with or interpret the environment? Does it have a knowledge structure it uses effectively and meaningfully? The chameleon is consigned to the dictates of whatever environment it finds itself in. Should this be the case with science and technology learning? It should not and there must be ways of integrating the two world-views. Both are valid, but could be strengthened by using factors that are congruent in both world-views to explain science to learners (see figure 10.1C).

Science should be taught and learned using all aspects of human endeavour — epistemological, technological, artistic, societal, cultural, private, prior or historical. Pomeroy (1994) has classified the agendas that address issues of cultural diversity and science in multicultural societies, and Aikenhead (1996) has proposed a theoretical framework for 'border crossing' from the subcultures of learners' peers and families to the subcultures of science and school science. Those faced with a duality of cultures in science classrooms would be better prepared if their experiences were structured so they could move from parallel collateral learning to secured collateral learning.

I am emphatic that collateral learning and language games are distinct. World language games are represented by figure 10.2 that looks at two separate worlds and two separate sciences. Collateral learning progresses beyond world language games in that the two worlds eventually merge. Learners move from: (1) constructing incompatible ideas in their mental schema from two worlds (parallel collateral learning), through (2) learning ideas from two worlds at the same time (simultaneous collateral learning), and (3) using ideas from one world-view to challenge or understand the views from another (dependent collateral learning) and, finally, learners (4) resolve cognitive conflict and convergence towards communality (secured collateral learning).

 

**Figure 10.2: Hierarchical organization of knowledge**

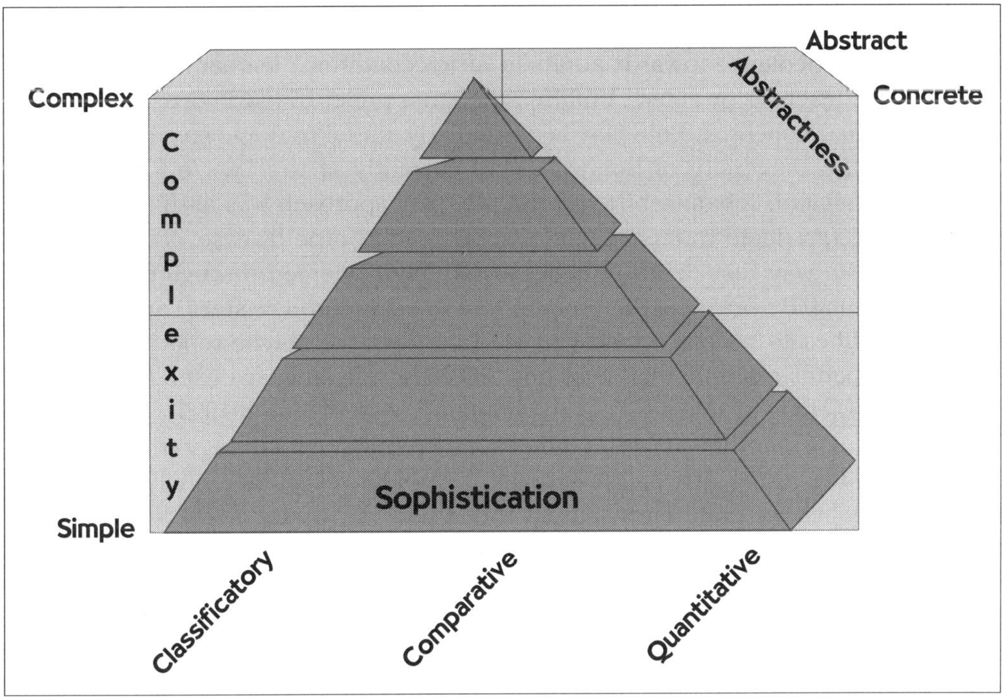


## INNOVATIVE TEACHING STRATEGIES FOR NON-WESTERN CLASSROOMS

In the West, innovative teaching strategies have been credited to scholars such as Piaget, Vygotsky, Freinet, Montessori and Dewey. Considering the contemporary understanding of knowledge structures and learning, there is a move towards analysing expert teachers to help explicate effective teaching (see Stenberg and Hovarth, 1995). Their instructional strategies have been copied and used in classrooms in Africa. Implementing curricula based on Western cultures requires Western instructional strategies. However, the design and management of our classrooms preclude a vigorous use of teaching strategies based on a variety of world-views. Colonial indoctrination that what is imported is best still seems to dominate in much of Africa. While in 1996 the West celebrates the centennial anniversary of the founding of Dewey's laboratory school in Chicago with a conference on innovative teaching, Africa should examine what has gone wrong with instruction in its own classrooms. Instruction is at the heart of implementing a curriculum. However well designed, if the content of a curriculum is not effectively communicated, efforts to build the curriculum remain ineffectual.

What effective instructional strategies do African cultures have that could be brought into the classroom? There are several used at home and in school outside the classroom. They include role playing, story telling, songs and dance, ceremonies, and rituals. I shall briefly mention two. Role play is common in African communities, be it during children's play, in open theatres, local festivities, or within the home and extended family. Role play enables children to appreciate what others are communicating and allows them to express their feelings indirectly. Respect for authority and unquestioned obedience towards adults in Africa conditions learners to avoid direct challenges to teachers or elders. Children therefore prefer to use indirect methods of expressing their concerns. Role play is particularly suited to doing so. Michael Kahn, formerly of the University of Botswana and currently at the Centre for Education Policy Development, Johannesburg, after a role play approach was used in teaching a group of undergraduate trainee teachers about water-borne disease, concluded that '[t]here was no way that depth of belief would have emerged through conventional science teaching. By breaking the boundary of the standard (Western) approach, the trainee was freed to become congruent with her own (or maybe reported) feelings. There is no better example I know of how effective free language expression can be used in the teaching of science' (e-mail communication of March 1995).

Story telling is another powerful indigenous instructional strategy that should be used more often in the African classrooms. I can still recall, as a child, waiting for night to fall when we children would sit round a fire or under a tree in the light of the full moon to listen to storytellers who used language and actions to evoke feelings and emotions. Such stories sounded so real to us that during the day we acted them out, uninterrupted by adults or elder siblings. More important, during story time we shared our feelings, experiences, thoughts, and what we learned in school and on the farm. On reflection, I now recognize how powerful this medium is in negotiating meaning. Unfortunately, structured classroom lessons, interrupted by bells and other constraints, rarely allow teachers the time or flexibility to use story telling. Martin and Bouwer (1991: 708), arguing for the need to use story telling in communicating science, stress that 'stories are our natural means of sharing in the lives of others and more fully exploring meaning in our own. Through stories, students may more successfully begin to see the subtle dimensions of science and of understanding the ways in [which] science, culture, and world-view interact'. Driver, Guesne, and Tiberghien (1985) also suggest story telling to help students explicitly formulate their own ideas, so that they are exposed to the contrast between their own perceptions and the conceptions offered by school science. The constructivist model is interested in how students personally make meaning in science and technology classes and views story telling as a powerful metacognitive pedagogical tool.

Most African countries have a legacy of colonial education that they have tried to reform. Many reforms failed because they were either 'panel beatings' of the old system or a substitution of one type of Western system of education for another. These foreign educational systems are not in themselves ineffective. They fail in Africa because they were designed to solve specific educational problems of the home environment, and are based on a Western world-view for those living within Western cultures. If attempts to graft foreign educational systems onto the African environment are not achieving the desired results, they cannot be compatible with the African environment. Perhaps Africa should critically examine traditional African educational systems and use their more powerful features as scaffolding for new educational systems that could carry aspects of the imported systems already in use. Some African scholars (for example Boateng, 1985) strongly believe that the introduction of Western education systems to Africa has seriously impeded the growth of the continent, especially in education, and has alienated Western-educated Africans from commitment to the development of indigenous values.

Indigenous African education systems incorporate imitation methods, initiation into age grades, and an apprenticeship system that Majasan (1976) has called the 'arduous training in specialized art and crafts'. The indigenous education system focused on learning and teaching that: (1) was related to the background of the learner; (2) was practical and involved learner participation; (3) incorporated local ideas and examples, using material resources within the immediate environment; (4) took place anytime, anywhere, anyhow and with due consideration to seeing the environment in holistic terms; and (5) used all competent people within the community as teachers and instructors. The recent use of a combination of the formal and indigenous systems of education in a correspondence mathematics course to vocational students (Akinlua, 1995) showed that, if given a chance, the indigenous system of education can be a positive and potent force for change in teaching science and technology in Africa.

## ORAL CULTURE, THE RURAL ENVIRONMENT, AND SCIENCE AND TECHNOLOGY

An oral culture and a rural population are central features of African societies. Many people living in rural areas lack access to Western education, are illiterate, and communicate orally. Some mistakenly think that to be educated means to be literate, though using local criteria, many people in rural areas who can only communicate orally are regarded as being highly educated. For instance many villagers are poetic, use idioms, proverbs, metaphors and analogies, and hold massive amounts of information in their mental schema. This is the basis of the sage system in Africa and other non-Western societies. Elders are ascribed the role of the all-wise and carry knowledge of the community that they pass on to the younger generation. Knowledge is power and the sage system concentrates power in the elder. Youth is denied this knowledge except as determined by elders through ceremonies, initiations, and so on. While rural areas still retain traditional cultures, urban areas are fast losing contact as they embrace Western cultures. The dilemma is how to strike a balance between the acquisition of foreign culture and those elements of the local culture to be retained. For guidance, we should look to the rural areas.

Harnessing the strengths of the rural and oral culture may be yet another panacea for the current problems of science and technology education in Africa. Hodson (1992) has cautioned against using a hegemonic hierarchy of knowledge to deny a study of science to a good percentage of people. Science and technology in schools emphasize abstract literacy-based skills that move learners away from relating their knowledge to real-life situations. The lack of value given orality in society and in science and technology classrooms is oppressive and selectively denies rural dwellers the opportunity to study science and technology fully. That the majority of our youth live in rural areas means that Africa is denied the full contributions that potentially good science and technology students could make to development. Rampal (1992), using the Indian experience as a reference point, is concerned about the relegation of the predominantly oral universe of children, especially those from nonliterate backgrounds, to what she qualified as a 'distinctly degraded subaltern status'. She accused countries of the Third World, where orality is still a major force in communication, of paying no attention to 'processes embedded in their own space of cultural and social cognition. They have instead denounced orality, tainting it with malice while christening it "illiteracy", officially proclaimed as an abhorrent affliction demanding determined eradication' (1992: 239). This certainly applies to every country in Africa. African science and technology educators should recognize the cognitive resources that abound in the linguistic structures and sociocultural environment of rural Africa and use them to chart a new course for the 21st century.

## Addressing gender inequity

There are more females than males in Africa. If, as stated by the United Nations Charter of 1948, education is the right of every citizen (they mean Western education), there should be equitable treatment of girls and women in Africa. Women and girls are underrepresented in the science and technology professions, careers, and educational institutions, whereas more women than men are farmers, traders, teachers and nurses. The contribution of women is crucial if Africa is to compete favourably in the world economy. The continent needs well-trained women scientists and technologists. Women are the educators of the young and their positive attitude to, and enhanced achievement in, science and technology would boost the participation of children, especially girls.

The perception of the role and status of women in African societies must change. Women must strengthen their self-image and see themselves as important contributors to society. To date, many African communities have seen the education of girls as secondary to that of boys. Teenage girls more often than not marry and receive no further science and technology education. However, though necessary, legislation rarely changes culture and tradition. A gradual change of attitude through education is required, as well as successful African women role models in science and technology (Harding & Apea, 1990). Rather than justifying intervention programmes for girls by using a deficit model (Atwater, 1994), perhaps we should examine other solutions. According to Pollina (1995), in the past we have focused on girls and women as if they are the problem. She says we often ask the questions, 'What is wrong with them, and how do we fix them?', or 'How do we make them more aggressive, more analytical, more competitive, tougher, so that they will survive in these disciplines?' (1995: 30). She recommends that instead of trying to change how girls approach science and technology, we need to study how they learn. Our research study into students' preferred and perceived classroom environment (Jegede, Agholor, &

Okebukola, 1996) revealed that girls prefer a less authoritarian classroom environment than boys, and classrooms where goal structures are evident and an emphasis is placed on group cooperation and collaboration rather than on individual competitive learning. An exploratory study which used Glynn's teaching-by-analogy model to investigate the use of sociocultural analogies in teaching biological concepts indicated that treatment corrected gender differences (Lagoke, Jegede & Oyebanji, 1996). We should look more closely at how girls learn, and use the African world-view more frequently as a basis for teaching science and technology.

## CONCLUSION

African countries have made little progress in teaching science and technology or in using science and technology for national development. Effective science and technology education is needed to develop an appropriate knowledge base. The main thesis of this chapter is that the African world-view needs to be central to future development and implementation of a new science and technology education for African schools. Science and everyday thinking should not be qualitatively different and collateral learning could solve problems of cognitive apartheid amongst African learners. Mention has been made of using indigenous innovative instructional practices, orality and African ruralness, and the African world-view.

African nations should review their science and technology education policies. They should be written within an African world-view and must include:

- the use of students' prior knowledge;
- content grounded in the child's immediate experience;
- the use of traditional instructional strategies;
- the presentation of science and technology concepts based on examples within the indigenous culture; and
- the promotion of a harmonious coexistence of world-views and their use to reinforce one other in concepts being taught.

Africa must reflect on the dual culture sweeping the continent to solve educational problems and move towards world-class achievements in science and technology. Although Western cultures came during the slave trade and colonial era, they have become incorporated into the fabric of our society. We expect our youth to participate in global developments and conform to the modern world, yet press them not to forget their indigenous culture. We must reach a balance that draws from the best of both cultures. I conclude that the road towards the emancipation of science and technology education in Africa will be long and rough. I am optimistic because there are many within the continent who are seeking realistic solutions and adopting unconventional approaches to science and technology education.

## ACKNOWLEDGMENT

I am indebted to Ivan Williams, Director of the College for Higher Education Studies, Suva, Fiji, for sharing ideas with me and reading through the first draft of this paper;

 

and to Professor and Mrs Philip Morrison of Cambridge, Ma, USA, for their useful comments, many of which are incorporated in this revision. All shortcomings are, however, mine.

## REFERENCES

Abimbola, IO. 1977. African world-view and school science. *Journal of the Science Teachers Association of Nigeria,* 16(1), pp 15–28

Aikenhead, GS. 1996. Science education: Border crossing into the subculture of science. *Studies in Science Eduction,* 26, (forthcoming)

Akinlua, AA. 1995. Effectiveness of a mathematics correspondence instruction on the vocational skills of junior secondary school graduate apprentices in Ondo State of Nigeria. Unpublished doctoral thesis, Obafemi Awolowo University, Ife, Nigeria

Atwater, MM. 1994. Research on cultural diversity in the classroom. In Gabel, DL (ed). *Handbook of Research on Science Teaching and Learning,* pp 558–78. New York: Macmillan Publishing Company

Baker, D & Leary, R. 1995. Letting girls speak out about science. *Journal of Research in Science Teaching,* 32(1), pp 3–27

Bamgbose, A. 1984. Mother tongue medium and scholastic attainment in Nigeria. *Prospects,* XIV(1), pp 87–93

Biesheuval, S. 1972. The ability of African children to assimilate the teaching of science. In Gilbert, PGS & Lovegrove, MN (eds). *Science Education in Africa.* Heinemann, London

Boateng, F. 1985. African traditional education. *Journal of Black Studies,* 13(3), pp 321–36

Brown, SJ, Collins, A & Duguid, P. 1989. Situated cognition and the culture of learning. *Educational Researcher,* 17(1), pp 32–42

Catsambis, S. 1995. Gender, race, ethnicity, and science education in the middle grades. *Journal of Research in Science Teaching,* 32(3), pp 243–57

Cobb, P. 1989. Experiential, cognitive, and anthropological perspectives in mathematics education. *For the Learning of Mathematics,* 9(2), pp 32–42

Cobern, WW. 1993. Contextual constructivism: The impact of culture on the learning and teaching of science. In Tobin, K (ed). *The Practice of Constructivism in Science Education,* pp 51–69. Washington, DC: AA Press

Cobern, WW. 1994, March. World-view theory and conceptual change in science education. Paper presented at the annual meeting of the National Association for Research in Science Teaching, Anaheim, Ca

Cobern, WW. 1995a, April. Everyday thoughts about nature: An interpretive study of 16 Ninth Graders' conceptualisation of nature. Paper presented at the annual meeting of the National Association for Research in Science Teaching, San Francisco, Ca

Cobern, WW. 1995b. Electronic message on world-view language games. Message no 01HV83-IZQ3OY8Y86VO@asu.edu of Wednesday, 13 September 1995 10:40AM

Collison, GO. 1974. Concept formation in a second language: A study of Ghanaian school children. *Harvard Educational Review,* 44(3), pp 441–57

Connelly, FM & Clandinin, DJ. 1990. Stories of experience and narrative inquiry. *Educational Researcher,* 19(5), pp 2–14

Cossons, N. 1993. Let us take science into our culture. *Interdisciplinary Science Reviews,* 18(4), pp 337–42

Driver, R. 1979. Cultural diversity and the teaching of science. In Trueba, H & Barrett-Mizrahi, C (eds). *Bilingual Multicultural Education and the Classroom Teacher: From Theory to Practice.* Rowley, Ma: Newbury

Driver, R, Asoko, H, Leach, J, Mortimer, E & Scott, P. 1994. Constructing scientific knowledge in the classroom. *Educational Researcher,* 23(7), pp 5–12

Driver, R & Easley, J. 1978. Pupils and paradigms: A review of literature related to concept development in adolescent science students. *Studies in Science Education*, 5, pp 61–84

Driver, R & Erickson, G. 1983. Theories-in-action: Some theoretical and empirical issues in the study of students' conceptual frameworks in science. *Studies in Science Education*, 10, pp 37–60

Driver, R, Guesne, E & Tiberghien, A. 1985. *Children's Ideas in Science.* Milton Keynes: Open University Press

Forde, D. 1954. *African Worlds: Studies in the Cosmological Ideas and Social Values of African Peoples.* London: Oxford University Press

Gagne, RM. 1975. *The Conditions of Learning*, 3 ed. New York: Holt, Rinehart and Winston

Gallagher, JJ & Dawson, D (eds). 1988. *Science Education and Cultural Environment in the Americas.* Washington, DC: NSTA, NIF, OAS

Glaser, R. 1991. The maturing of the relationship between the science of learning and cognition and educational practice. *Learning and Instruction*, 1, pp 129–44

Harding, J & Apea, E. 1990. *Women Too in Science and Technology in Africa: A Resource Book for Counselling Girls and Young Women.* London: Commonwealth Secretariat

Hedges, LV & Waddington, T. 1993. From evidence to knowledge to policy: Research synthesis for policy formation. *Review of Educational Research,* 63(3), pp 245–47

Hodson, D. 1992. Towards a framework for multicultural education. *Curriculum,* 13(1), pp 15–18.

Idowu, EB. 1963. *African Traditional Religion: A Definition.* London: SCM Press Ltd

Jegede, OJ. 1995a. Collateral learning and the eco-cultural paradigm in science and mathematics education in Africa. *Studies in Science Education*, 25, pp 97–137

Jegede, OJ. 1995b. School science and the development of scientific culture: A review of contemporary science education in Africa. *International Journal of Science Education* (in press)

Jegede, OJ. 1995c. Response of Thursday, 14 September 1995 12:12PM to Bill Cobern's electronic message on world-view language games. Message no 01HV83IZQ3OY8Y86VO-@asu.edu of Wednesday, 13 September 1995 10:40AM

Jegede, OJ. 1994. African cultural perspectives and the teaching of science. In Solomon, J & Aikenhead, G (eds). *Science Technology Society Education for Future Citizens,* pp 120–30. New York: Teachers College Press

Jegede, OJ. 1989. Toward a philosophical basis for science education of the 1990s: An African view point. In Herget, DE (ed). *The History and Philosophy of Science in Science Teaching,* pp 185–98. Tallahassee: Florida State University

 

Jegede, OJ, Agholor, R & Okebukola, PAO. 1996. Does gender make a difference? A study of the perceived and preferred socio-cultural science classroom climate in Nigeria. *Research in Education* (forthcoming)

Jegede, OJ, Fraser, BJ & Okebukola, PAO. 1994. Altering socio-cultural beliefs hindering the learning of science. *Instructional Science*, 22, pp 137–52

Jegede, OJ & Okebukola, PAO. 1993. Measuring the effects of socio-cultural factors in non-western science classrooms. *The Hong Kong Journal of Educational Research*, 8, pp 40–7

Jegede, OJ & Okebukola, PAO. 1992. Differences in socio-cultural environment perceptions associated with gender in science classrooms. *Journal of Research in Science Teaching,* 29(7), pp 635–47

Jegede, OJ & Okebukola, PAO. 1991. The effect of instruction on socio-cultural beliefs hindering the learning of science. *Journal of Research in Science Teaching*, 28(3), pp 275–85

Jegede, OJ & Okebukola, PAO. 1990. The relationship between African traditional cosmology and students' acquisition of a science process skill. *International Journal of Science Education,* 13(1), pp 37–47

Jegede, OJ & Okebukola, PAO. 1989. Some socio-cultural factors militating against drift towards science and technology in secondary schools. *Research in Science and Technological Education*, 7(2), pp 141–51

Jegede, OJ & Okebukola, PAO. 1988. The educology of socio-cultural factors in science classrooms. *International Journal of Educology*, 2(2), pp 93–107

Jegede, OJ & Olajide, JO. 1995. Wait-time, classroom discourse and the influence of socio-cultural factors in science teaching in the Nigerian context. *Science Education*, 79(3), pp 233–49

Jegede, OJ, Naidoo, P & Okebukola, PAO. 1996. The validity of the science student stress inventory using a sample of South African high school students. *Research in Science and Technological Education,* 14(1) Carfax Publishing Ltd.

Jung, W. 1993. Uses of cognitive science to education. *Science and Education*, 2(1), 31–6

Kahle, JB & Meece, J. 1994. Research on gender issues in the classroom. In Gable, D (ed). *Handbook of Research on Science Teaching and Learning,* pp 542–57. New York: Macmillan Publishing Company

Kant, I. 1965. *Critique of Pure Reason* (NK Smith, trans). New York: St Martin's Press. (Original work published in 1781)

Kelly, J. 1955. *The Psychology of Personal Constructs.* New York: Norton

Kerderman, D & Phillips, DC. 1993. Empiricism and the knowledge base of educational practice. *Review of Educational Research,* 63(3), pp 305–13

Lagoke, BA, Jegede, OJ & Oyebanji, PK. 1996, January. Towards eliminating the gender gulf in science concept attainment through the use of environmental analogs. Paper prepared for the 8th International Gender and Science and Technology Conference (GASAT 8), Ahmedabad, Gujarat, India

Loving, CC. 1995. Comment on multiculturalism, universalism, and science education. *Science Education,* 79(3), pp 341–48

 

Majasan, JA. 1976. Traditional system of education in Nigeria. *Nigerian Magazine*, 119, pp 23–9

Martin, BE & Bouwer, W. 1991. The sharing of personal science and the narrative element in science education. *Science Education,* 75(6), pp 707–22

Mbiti, JS. 1969. *African Religions and Philosophy.* London: Heinemann Educational Books Ltd

Ogawa, M. 1986. Towards a new rationale of science education in a non-Western society. *European Journal of Science Education*, 82, pp 113–19

Ogawa, M. 1995a. Science education in a multi-science perspective. *Science Education,* 79(5), pp 583–93

Ogawa, M. 1995b. Four-eyed fish: The ideal for non-Western graduates of Western science education graduate programmes. *Science Education,* 79(6), in press

Ogbu, JU. 1992. Understanding cultural diversity and learning. *Educational Researcher,* 21(8), pp 5–14, 24

Ogunniyi, MB. 1987. Conceptions of traditional cosmological ideas among literate and non-literate Nigerians. *Journal of Research in Science Teaching,* 24(2), pp 107–17

Ogunniyi, MB. 1988a. Adapting western science to traditional African culture. *International Journal of Science Education,* 10, pp 1–9

Ogunniyi, MB. 1988b, September. Sustaining students' interests in science and technology: The socio-cultural factors. Lead paper presented at the 29th Annual Conference of the Science Teachers' Association of Nigeria, Ibadan

Ogunniyi, MB, Jegede, OJ, Ogawa, M & Yandilla, C. 1995. World views projected by science teachers in Nigeria, Japan, Botswana, Indonesia and the Philippines. *Journal of Research in Science Teaching,* in press

Okebukola, PAO & Jegede, OJ. 1990. Eco-cultural influences upon students' concept attainment in science. *Journal of Research in Science Teaching,* 27(7), pp 661–69

Osborne, J, Bell, B & Gilbert, J. 1983. Science teaching and children's views on the world. *European Journal of Science Education,* 5(1), pp 1–14

Phelan, P, Davidson, A & Cao, H. 1991. Students' multiple worlds: Negotiating the boundaries of family, peer, and school culture. *Anthropology and Education Quarterly,* 22(3), pp 224–50

Piaget, J. 1970. *Science of Education and Psycology of the Child.* New York: Orion Press

Pollina, A. 1995. Gender balance: Lessons from girls in science and mathematics. *Educational Leadership,* 53(1), pp 30–3

Pomeroy, D. 1994. Science education and cultural diversity: Mapping the field. *Studies in Science Education,* 24, pp 49–73

Prawat, RS. 1993. The values of ideas: problems versus possibilities in learning. *Educational Researcher,* 22(3), pp 5–16

Prophet, RB. 1990. Rhetoric and reality in science curriculum development in Botswana. *International Journal of Science Education,* 12(1), pp 13–23

Resnick, LB. (ed). 1989. *Knowing, Learning, and Instruction. Essays in Honour of Robert Glaser.* Hillsdale, NJ: Lawrence Erlbaum Associates

Rampal, A. 1992. A possible orality for science. *Interchange,* 23(3), pp 227–44

Rampal, A. 1994. Innovative science teaching in rural schools in India: Questioning social beliefs and superstition. In Solomon, J & Aikenhead, G (eds). *Science Technology Society Education for Future Citizens,* pp 131–38. New York: Teachers College Press

Science Teachers' Association of Nigeria (STAN). 1992. Women in science, technology and mathematics: The Nigerian experience. Position Paper 2. Ibadan: Science Teachers' Association of Nigeria

Snively, G. 1989. Traditional native Indian beliefs, cultural values and science instruction. *Canadian Journal of Native Education,* 17(1), pp 44–59

Solomon, J. 1989. Social influences on the construction of pupils' understanding of science. *Studies in Science Education,* 14, pp 63–82

Solomon, J. 1994. The rise and fall of construtivism. *Studies in Science Education,* 23, pp 1–19

Stenberg, RJ & Hovarth, JA. 1995. A prototype view of expert teaching. *Educational Researcher,* 24(6), pp 9–17

Swift, D. 1992. Indigenous knowledge in the service of science and technology in developing countries. *Studies in Science Education,* 20, pp 1–28

Thomas, RG, 1992. *Cognitive theory-based teaching and learning in vocational education.* Information series No 349. Columbus: ERIC Clearing House on Adult, Career, and Vocational Education (ED 345109)

Tindimubona, A. 1991. Science culture in Africa. *Whydah — African Academy of Sciences Newsletter,* 2(6), pp 14

Tobin, K. 1995. Message posted on his behalf by Alejandro Gallard, on RESODLAA online discussion list of Saterday, 23 September 95 18:21:58 EST, Message-Id: v01530500-ac8839bb5563@[128.186.11.158]

Vygotsky, LS. 1978. *Mind in Society: The Development of Higher Psychological Processes.* Cambridge, Ma: Harvard University Press

Von Glasersfeld, E. 1989. Cognition, construction of knowledge and teaching. *Synthese,* pp 121–40

Wang, MC, Haertel, GD, & Walberg, HJ. 1993. Towards a knowledge base for school learning. *Review of Educational Research,* 63(3), pp 249–94

Wheatley, GH. 1991. Constructivist perspectives on science and mathematics learning. *Science Education,* 75(1), pp 9–21

Wertsch, JV & Toma, C. 1992. Discourse and learning in the classroom: A sociocultural approach. In Steffe L, (ed). *Constructivism in Education.* Hillsdale, NJ: Lawrence Erlbaum Associates

Wiredu, K. 1980. *Philosophy and an African Culture.* Cambridge, Ma: Cambridge University Press

# 11

## Research in science and technology education

*Prem Naidoo, University of Durban-Westville, Durban, South Africa*

### ABSTRACT

This chapter surveys science education research in selected Anglophone countries in East, West and southern sub-Saharan Africa. It: (1) reviews literature on the state of educational and science education research; (2) analyses some science education research publications; (3) presents an analysis of the responses to a questionnaire intended to survey researchers in Ghana, Malawi, Nigeria, Swaziland, Uganda and Zimbabwe; and (4) gives the findings of interviews conducted with 10 African science educators.

### LITERATURE REVIEW

The review helped construct an analytical framework to define educational research and determine factors that may influence science education research.

Court (1983) sees research as a systematic production of knowledge about the functioning and impact of any system. Analysis of this definition suggests that educational research involves three basic elements (Keeves, 1990), namely, the creation of knowledge, the use of the knowledge by policy makers and practitioners, and the diffusion of knowledge through mechanisms that link the creation of knowledge with its use.

Educational research should unify the creation, diffusion and use of knowledge. (Husen, 1990; Tipane, 1990). Creation presupposes use, and without appropriate mechanisms for diffusion and dissemination, creation and use remain unlinked. Moreover, failure to recognize that knowledge of educational processes should change policy and practice has given rise to criticisms that educational research is an ineffective and inappropriate tool for promoting change (Keeves, 1990).

There are many models of educational research and a variety of research designs, methods and processes (Keeves, 1990; Walker, 1990). Generally, researchers use quantitative methods, qualitative methods, or a combination of the two. An important question is whether Africa can afford research that does not emphasize all three equally.

The main purpose of research is to understand and improve social conditions and institutions. Educational research should uncover information that can be used by a range of people, from national policy makers to classroom teachers, so that they can improve equity and quality of learning. Research that contains all three basic elements in a balanced way would do so, and it is urgent that academics engage more in such work. Research is done within a social setting consisting of related components that may affect its type, quality and impact (Schaeffer & Nkinyangi, 1983; Court, 1983; Keeves, 1990). To understand the state of research in Africa better, researchers must seek answers relating to these components that would include: (1) the education system; (2) the state and nature of the system and its affect on research; (3) the research climate or culture; (4) research processes, skills and competencies; (5) research infrastructure and support; (6) research performance; (7) donor involvement, and (8) the role of government.

This chapter asks questions on each component, and the research methodologies used were chosen to provide the information required to answer them.

## Education research

Educational research in sub-Saharan Africa is periodically reviewed by donor organizations such as the International Bank of Reconstruction and Development (IBRD) in 1980; the International Development Research Council (IDRC) in 1983 and 1991 (Schaeffer and Nkinyangi); and UNESCO in 1990 (Yoloye). Other studies include those by Court (1982, 1983, and 1991), Evans (1994), the African Academy of Sciences (1992), UNESCO (1994 and 1995), Sherman (1990), Hallak and Fagerling (1991), and others.

All reviews acknowledge that strengthening research and analytical capacities is an essential requirement for the improvement of educational systems. Educational policies and practices, as well as decisions taken regarding them, must be informed by the results of systematic, well-conceived research, evaluation and assessment. Reportedly, progress has been made in Africa in achieving this during the last decade. The number of researchers, research institutions and research training programmes has increased. Some have become focal points for dynamic research, often in collaboration with national, regional and international networks. Donors are realizing the importance of the development of local research capacities, of more flexible training models, and of sustaining the research and analysis process. In some countries, research has become more valued than in the past.

According to the literature review, however, more must be done to improve African educational research. The author identified major problems that were referred to by most studies reviewed. These include problems of:

- Declining economies, with concomitant cuts in funding for research that have led to deteriorating infrastructures and hindered the growth of healthy research cultures. As perceived by reviewers, research is not seen as a priority in Africa, in the same way as the provision of basic education and primary health is underrated.

- Authoritarian governments suppressing research and distrusting intellectuals. When they fund research, work is inspired by the politics of the ruling party and frequently has little to offer other than showing the measures taken to promote equity and excellence.
- Inadequate data and information bases in some countries on the size, quality and costs of education. In other countries, there may be sufficient data that is inadequately selected, analysed and presented for easy use in decision making. Without accurately collected and rapidly processed data (the studies reviewed claim), policy analysis and decision making will remain ill informed.
- A lack of human resources in some countries. Others often underuse their own researchers, preferring the expatriate consultant — thereby removing support from local institutions.
- Research paradigms, frameworks and methods being inappropriate to the educational problems that face Africa. The studies reviewed reveal that most approaches to collecting and analysing data evolved in developed countries and may require modification when used in Africa. Such techniques focus on problems, do not suggest solutions, and rarely involve policy studies, qualitative methodologies or action research. The dominance of research by donors is reported as leading increasingly to the setting of research priorities rather than this being done by national institutions.
- A weak demand for educational research because parents may not have been concerned with the quality of education received by their children. Education research, instead, was driven by donor needs, the donors being more concerned with feasibility studies and evaluations of their own projects.
- Minimal networking between African researchers that limits professionalism. The studies noted that it is easier for professionals to meet outside Africa than within the continent. Limited or erratic post and telecommunications make communications difficult and the community of electronic networkers in Africa is small.
- A lack of regularly published journals, limiting professional growth.
- Narrow and inappropriate research agendas what fail to address educational problems in Africa. The studies reviewed noted that most research in Africa is for higher degrees and there is little concern for the use of results towards an improvement of the system.
- Research being almost entirely based at universities, with little involvement from other institutions within education.
- A focus on research products rather than on the process, which has reportedly led to poor work and deteriorating infrastructures.
- Poor working conditions for researchers. This has resulted in their being faced with increased classes and workloads. Poor salaries, the studies claim, lead to skilled researchers being forced to seek other jobs to supplement their incomes.

- Inappropriate training of researchers, generally in institutions outside Africa. It is claimed that they find it difficult to work in the conditions they encounter when they return home. Furthermore, their research agendas are frequently influenced by the institution at which they studied, rather than by local concerns. National research institutions are finding it increasingly difficult to provide adequate training as senior, experienced mentors leave to join NGOs, international agencies, or the donor community.

All educational research must be critiqued within a context of knowledge, power, cultural struggle and possibility, rather than using analytical frameworks that perpetuate dominant ideologies. Central to research is the development of an articulated language of possibility and critical competencies necessary to reveal and deconstruct all forms of oppression. Since much research focuses inquiry on education (a moral action), rather than for and in the service of education (an ethical action), Kyle (1995) claims it is not surprising that there are few examples of how teaching has been improved through research. Certainly it is not surprising that such surveys as do take place in highly detached settings far removed from the context of the lived experience of teachers and learners should have a negligible impact. Such 'traditional positivistic and interpretive epistemological perspectives to research,' Kyle claims, 'offer little to those wishing to improve schooling' (1995: 6).

Namuddu (1991a and 1991b) also contests the findings of most international reviews of educational research in Africa. She argues that voices within and without Africa contain persistently negative messages and have exploited the ideology of poverty by repeated claims such as were detected in the literature review done during this study. Namuddu summarizes the claims frequently made by such international reviews, which state that Africa:

- Lacks an adequate research capacity in education and policy analysis.
- Does not have the capacity to preside over reforms in education.
- Produces educational research of low quality.
- Lacks adequately trained and experienced personnel to plan, manage and administer educational research institutes and other educational institutions.
- Produces little or no research on major structural, organizational and policy initiatives in African educational systems.
- Does not make research results available when required.
- Produces research that tells us little or nothing about what is happening in African education.

She argues that data reported in internationally sponsored surveys is frequently collected using questionnaires designed by the same agencies. Those responsible for administering the research instruments attempt to fit information to the pattern predetermined by the donor. Often the purpose of these surveys is to make international comparisons. They are designed with the industrial North in mind, and distort the African perspective.

Namuddu stresses that to obtain a more balanced understanding of educational research in Africa, it is necessary to access information from a perspective that would involve questioning: (1) whether the research is inspired by national or foreign influences; (2) the motives underlying research claims; and (3) how the research has been affected by historical influences such as the colonial experience.

Her analysis, based on such questioning, leads her to argue that foreign-inspired research seeks to perpetuate a philosophy and organizational pattern which insists that Africa is lacking or weak in components such as those identified in the literature review conducted in this study. She would add the following to the list of foreign-inspired research claims: (1) a lack of appropriate instructional resources and facilities; (2) inadequately trained and poorly motivated teachers; (3) poorly designed assessment systems, and (4) insufficient trained planners and managers. A common recommendation, therefore, of internationally inspired surveys, Namuddu concludes, is that such defects be met by importing related educational models and expertise from the North.

Namuddu feels that whoever conducts research is linked to issues of motive and funding. The research climate and patterns in Africa were established by expatriates at universities and ministries. They used adequate budgets to develop and sustain research and supporting infrastructures and to set research priorities, and used their home institutions rather than local ones as their reference group. Africans were junior staff members who acted as research assistants. The situation changed after independence, partly because of staff changes in research institutions, and partly because of the changing sociopolitical climate at national and international levels. National governments needed funds for the purpose of expanding of access at all educational levels, for implementing crash programmes to train teachers and administrators, for developing localized curriculum materials and so on. Allocations to research were inevitably cut, not through a lack of appreciation for its role, but rather because of priorities within the fragile constituencies of nation states newly liberated from colonial rule. Government cuts to research were exacerbated as expatriates left, a need to 'Africanize' developed, and foundations such as Ford and Rockefeller, concerned with capacity building, invested substantially in training Africans overseas, leaving fewer funds to support research and research institutions within the continent. However, Namuddu claims that the most critical factor to influence the development, survival and invisibility of research in Africa is not funding, but the emerging social and political atmosphere. Education was perceived by national governments and politicians — abetted by donors — as the single most important instrument to promote equity and economic development, and by parents as a newly opened gateway to prosperity for their children. Faced with a clamour from the public and factions within their ranks, governments were eager to demonstrate their acumen for unification by displaying their ability to distribute resources fairly, and chose education for this purpose. Often educational policies have been decreed overnight and implemented haphazardly without consideration of how they would affect quality. The informed, more cautious voice of researchers increasingly becomes overwhelmed by the roar from various platforms praising leaders for their wise solutions to ills inherited from their colonial predecessors.

For such reasons, Namuddu opines that much indigenous research is fugitive. An implication is that few surveys can be expected to reveal a truly comprehensive picture. Methodological flaws detected in northern-inspired surveys constrain the building of a representative picture of African research — by definition they must be cartoons. Though she is highly critical of the use of positivistic paradigms used in such surveys, Namuddu has more sympathy with research methodologies that use qualitative and phenomenological perspectives which, in her judgement, present pictures that are more faithful to local practices.

## Science education research

This author found few surveys on or analyses of science education research in sub-Saharan Africa. Those surveys he identified concerned Nigerian science education research, conducted by Bajah (1990) and Obioma (1990). Two surveys of science education research in South Africa were conducted by Reddy (1995) and Lewin (1995).

The Nigerian reviews indicate that:

- Most science education researchers are located at universities, where most research is aimed at earning a higher degree or promotion.
- There is no shortage of university-trained researchers.
- Government funds university research, albeit inadequately.
- There are reasonable research infrastructures with access to journals, data bases, dataprocessing facilities and links with other national research institutes.
- There is an established research culture and tradition in science education.
- Research is focused largely on instructional materials and learner and teacher characteristics.
- Most research employs ex-post and quantitative methods.
- There is little action research on intervention strategies.
- Few longitudinal studies have been done.
- Most research is done by individual scholars; there is no tradition of collaborative work.
- There is little research on large classes, improvization, science education in rural areas, the role of national languages, or on policy.
- Research in science education is rarely directed by stated national needs.

South African reviews reveal that:

- Researchers are largely located at universities, where there are race and gender imbalances. For example, few South African science education researchers are black women.
- Research is funded by government and parastatals.
- Research tends to focus on cognition, particularly at secondary and tertiary levels.
- Little work is done on teacher education, environmental education, technology education, gender, equity and black access, financing, cost effectiveness, baseline studies, or policy and impact studies, particularly of NGOs.

- A qualitative research tradition dominates.
- Though only recently formed, the Southern African Association of Research in Mathematics and Science Education (SAARMSE) is playing a key role in the development of a strong research culture.
- Research infrastructures are well developed.

The study by Kahn and Rollnick, which reviewed science education research in West and southern Africa, analysed 80 science education research articles mainly from West and southern Africa. They were written by 103 authors, of whom 46 % were African. Forty percent of the articles appeared in journals, with 6 % appearing in African journals. Only 25 % of articles were written by two or more persons.

Analysis of the articles reviewed revealed that most focused on cognition, constructivism, African thought and ethnoscience. Little or no research was done on curriculum reconstruction, language of instruction, implementation and the impact of innovations, policy, or the relationship of science education to national development.

## ANALYSIS OF SCIENCE EDUCATION RESEARCH JOURNALS

There are few science education research journals in Africa. Journals in Ghana and Botswana closed after publication of only a few issues. SAARMSE hopes to publish a journal of science education research in southern Africa. The only science education journal in Africa that has been successfully published on a regular basis over many years is that of the Science Teachers' Association of Nigeria (STAN). This journal is peer reviewed and has an editorial board of Nigerians and a few outsiders. The contents are organized in sections that include one for articles of general interest, one for research reports, and a section for networking news and teaching notes. The journal is targeted at both researchers and teachers.

Four STAN journals published between 1991 and 1993 were analysed. Table 11.1 below presents the analysis of the section of articles of general interest and that of research reports.

The table strongly suggests that most articles are written by individual university researchers, confirming the analysis by Kahn and Rollnick (1994). There seems to be no tradition of cooperative research. Reasons may include that the requirements for higher degrees and promotion inculturate individual research. Positivistic and quantitative research predominates.

Table 11.2 below shows the areas that were researched. Analysis again confirms that of Khan and Rollnick (1994). It suggests that most research focuses on assessment and the learning characteristics of students. Little was done on topics such as teacher education, improvization, language of instruction, and science learning in rural schools. There seems to be silence on issues such as financing, equity, planning and policy — all research areas suggestive of Namuddu's 'fugitive' hypothesis. (1991b)

Over the five-year period, only five of 240 articles (3 %) published were from Africa. *Science Education*, however, cannot be blamed. Only 27 manuscripts were submitted from Africa during that period, of which seven were accepted — an acceptance rate of about 26 %. This compares favourably with an overall acceptance rate of 20 %.

**Table 11.1: Analysis of general-interest articles and research reports in STAN journals, 1991-93**

| Research areas | Number |
|---|---|
| Total number of articles analysed | 37 |
| Authors from departments of education | 1 |
| Authors from colleges of education | 7 |
| Authors from universities | 40 |
| Nigerian authors | 42 |
| Authors from outside Nigeria | 6 |
| Articles written independently | 27 |
| Articles written collaboratively | 10 |
| Articles using quantitative methodology | 31 |
| Articles using both quantitative and qualitative methodology | 1 |
| Articles that were purely literature reviews | 5 |

**Table 11.2: Analysis of areas researched in study of STAN journals, 1991-93**

| Research areas | Number |
|---|---|
| Rural schools | 1 |
| Improvization | 1 |
| Language instruction | 1 |
| Curriculum materials | 1 |
| Students success | 1 |
| Management and personnel of science education | 1 |
| Attitudes of learners to science | 2 |
| Teacher education | 3 |
| Cognition and student learning | 8 |
| Assessment and student outcomes | 17 |

Gibbs (1995) reports that some scientists from developing countries feel there is a bias on the part of international journals that results in the rejection of their manuscripts. To examine the situation in science education, the author approached two international journals, namely the *Journal of Research in Science Teaching* and *Science Education*. Only *Science Education* responded.

*Science Education* is published six times a year. Its contents are organized into a general section, a special section and a comments and criticisms section. The special section is subdivided into subsections. These currently are, science teacher education, learning, issues and trends, and international science education. The journal publishes on average eight articles per issue in these sections. Thus they publish approximately 48 per year, and would have published 240 during the period 1990–94.

**Table 11.3: Acceptance rates by *Science Education* of articles submitted from Africa between 1990 and 1994**

| Period 1990–94 | Number |
|---|---|
| Approximate total manuscripts received from all over the world | 1 200 |
| Overall % acceptance rate of articles from all over the world | 20 % |
| Total manuscripts received from Africa | 27 |
| Total manuscripts accepted from Africa for publication | 4 |
| Total rejected from Africa without review | 12 |
| Total rejected from Africa with review | 6 |
| Total revisions pending from African authors | 3 |

## ANALYSIS OF QUESTIONNAIRES

A questionnaire was developed and piloted in Uganda, then mailed to science education researchers in Botswana, Ghana, Lesotho, Malawi, Nigeria, Swaziland, Tanzania, Zambia and Zimbabwe. At least 10 questionnaires were sent to each country.

Only 36 questionnaires were returned: three each from Ghana and Malawi, four from Zimbabwe, five from Swaziland, eight from Nigeria and 13 from Uganda. One reason why the response rate was low may have been the length of the questionnaire. The response rate from Uganda was high because the author personally administered the questionnaire. All the respondents stated that primarily university academics are involved in research. Government education officers and teachers rarely do research, except as part of higher-degree programmes.

Table 11.4 below shows conditions experienced by academics that may influence their ability to do research.

**Table II.4: Analysis of 36 respondents' responses to questionnaire directed at science education researchers in Africa**

| Questions | Nigeria | Ghana | Uganda | Malawi | Zimbabwe | Swaziland |
| --- | --- | --- | --- | --- | --- | --- |
| **Gross monthly salary at lecturer level in US$** | 202 | 240 | 213 | 292 | 766 | 1 366 |
| **Who provides housing subsidy?** | University | University | University | University | Self, university provides housing allowance | Self, university provides housing allowance |
| **Do you have your own car?** | 75 % of respondents own cars | 67 % of respondents own cars | 30 % of respondents own cars | 50 % of respondents own cars | 50 % of respondents own cars | 100 % of respondents own cars |
| **Do you have access to computers?** | 100 % have access, 37,5 % can use a computer, and own a computer | 100 % have access, 33 % can use a computer, and own a computer | 54 % have access, 46 % can use a computer, and own a computer | 100 % have access, and can use a computer, and 57 % own a computer | 100 % have access, can use a computer and 25 % own a computer | 100 % own and can use a computer |
| **Cost of PC and printer in relation to monthly lecturer salary (assuming cost = US$2 000)** | 10 months' salary | 18,3 months' salary | 9,4 months' salary | 6,9 months' salary | 2,6 months' salary | 1,5 months' salary |

| Questions | Nigeria | Ghana | Uganda | Malawi | Zimbabwe | Swaziland |
|---|---|---|---|---|---|---|
| **Do you have access to research software?** | 62,5 % have access to SPSS | 67 % have access to SPSS | 7,6 % have access to SPSS and 23 % to Excel | 33 % have access to Statworks and Excel | 75 % have access to SPSS and 50 % to Lotus1–2–3 | 20 % have access to SPSS, Minitab, Excel and Lotus 1–2–3 |
| **Do you have access to data banks?** | 25 % have access to Eric and 50 % to govnt stats | 67 % have access to govnt stats | no-one has access to any data bank | 67 % have access to Eric and govnt stats | 100 % have access to Eric and to govnt stats | 80 % have access to Eric and to govnt stats |
| **How do you rate your collection of indigenous literature?** | 50 % found it to be good | all found it poor | 25 % found it to be good | 75 % found it to be satisfactory | 50 % found it to be good | 33 % found it to be good |
| **Do you have access to international literature?** | 37,5 % found it satisfactory | 37,5 % found it satisfactory | 50 % found it good | 67 % found it satisfactory | 50 % found it good | 67 % found it good |
| **Is communication (post & tel) easy within country?** | 87,5 % found post comm easy & 52% found tel comm easy | 33 % found both easy | 60 % found post comm easy & 40 % found tel comm easy | 100 % found post comm easy & 67 % found tel comm easy | 100 % found post comm easy & 75 % found tel comm easy | 80 % found post comm easy & 100 % found tel comm easy |
| **Can you communicate easily with a person outside your country?** | 87,5 % found post comm easy & 75 % found tel comm easy | 67 % found both easy | 84 % found post comm easy & 59 % found tel comm easy | 67 % found both easy | 100 % found post comm easy & 75 % found tel comm easy | 0 % found post comm easy & 100 % found tel comm easy |

| Questions | Nigeria | Ghana | Uganda | Malawi | Zimbabwe | Swaziland |
|---|---|---|---|---|---|---|
| **What % of lecturer salary will it cost to post one A4 letter and make one 3-min tel call within district, nationally and internationally?** | 9,7 % | 5,2 % | 4,6 % | 3,3 % | 0,86 % | 0,5 % |
| **Do you have e-mail access?** | No-one has access to e-mail | 67 % have access to e-mail | 15 % have access to e-mail | 100 % have access to e-mail | 75 % have access to e-mail | 100 % have access to e-mail |
| **Do you have access to a photocopier?** | 100 % have access to copier | 100 % have access to copier | 100 % have access to copier, but have to pay for photocopying | 100 % have access to copier, but have limited funds to pay for photocopying | 100 % have access to copier, but have limited funds to pay for photocopying | 100 % have access to copier and have access to funds to pay for photocopying |
| **Do you have access to a professional organization?** | Yes, to National Science Teacher and Educational Research Associations | Yes, to National Science Teacher Association | No | Yes, to National Science Teacher Association | Yes, to National Science Teacher and Educational Research Associations | Yes, to National Science Teacher, Educational Research and to Regional Science Education Research Associations |
| **On average how many hrs per wk do you teach?** | 11 | 11,5 | 8,25 | 6,3 | 4,67 | 7 |

| Questions | Nigeria | Ghana | Uganda | Malawi | Zimbabwe | Swaziland |
|---|---|---|---|---|---|---|
| **How many students do you teach per week** | 397 | 64 | 43 | 83 | 61 | 14 |
| **How many hours do you assign to research per week?** | 13,7 | 6,3 | 3,6 | 5 | 13 | 15,75 |
| **Do you have access to research funding?** | Limited funding from institution and nationally. Rely mainly on international funding | Reasonable access to funding from institution and internationally | Had no funding from institution till this year. Reasonable access to international donors | Poor institutional and national funding. Rely mainly on international funding | Limited access to institutional, national and international funding | Access to institutional funding and limited access to international donors |
| **Can you consult someone when doing research?** | Yes | Yes | Yes | Yes | Yes | – |
| **What is the highest qualification you have and from which country?** | 50 % each MEd and PhD. All education in Nigeria | 33 % MEd from Ghana and 67 % PhD from Nigeria and Canada | 23 % BEd from Tanzania, 87 % Masters from UK, Kenya and 23 % from Uganda | 33 % MSc from Germany and 67 % from South Africa | 25 % MA from US and 75 % PhD from US and Holland | 100 % Masters from UK, Holland and Canada |
| **Did you attend research methodology courses?** | All attended research methodology courses as part of formal study | All attended research methodology courses as part of formal study | All attended research methodology courses as part of formal study | All attended research methodology courses as part of formal study | All attended research methodology courses as part of formal study | All attended research methodology courses as part of formal study and at SAARMSE |

University research appears to be a well-established tradition. Most universities surveyed have university science education departments, most of which offer Masters' programmes. Over the last five years, 11 respondents collectively supervised 132 Masters and 19 PhD students. Approximately 70 % of these higher-degree candidates were in Ghana and Nigeria, which have had such programmes for many years. With the exception of Uganda, all countries surveyed have professional science teachers' associations. SAARMSE, a southern African organization, appears to be the only professional association for science education researchers in sub-Saharan Africa. There are few similar international bodies, an example being the National Association for Research in Science Teaching. A North American organization, this has an annual subscription of US$100, half the monthly salary of a Nigerian university lecturer. Research productivity was somewhat low. The 36 respondents conducted 61 studies between 1990 and 1994 — an average of about 1,5 years for a researcher to complete a study. Researchers surveyed used balanced research methodologies. About a third (22) of the studies used quantitative methodologies, about a fifth (11) qualitative, and close to half (28) used both. Other than research for higher degrees or promotion, most work done by the researchers surveyed appears to have been initiated by the funder. Close to 40 % (24) of the studies were self-funded. Most of these were initiated by the researcher to obtain a higher degree. The balance were funded and initiated by their institutions, or by publishers, international donor agencies and national research institutions. Only two studies were initiated and funded by government. On the whole, the analysis of questionnaires confirmed findings concerning areas of research identified by the literature review. Many studies focused on curriculum reform, instruction, teacher attitude, assessment and INSET. Few focused on policy, planning, language of instruction, improvization or gender — all controversial topics for politicians. There were no studies on ethnoscience, appropriate technology, human resource development, or teaching in resource-poor conditions. Perhaps these are viewed as interesting areas by the international research community, but not by nationals and donors.

## ANALYSIS OF INTERVIEWS

The author interviewed a total of 10 junior and seasoned education and science education researchers, using an unstructured approach. They were selected from Ghana, Kenya, Malawi, Nigeria, Uganda and Zimbabwe. All agreed that research was mainly done by university academics because of the enabling environment of these institutions. All remarked that conditions in universities had deteriorated, resulting in a decline in research productivity. Academics in countries that have experienced particularly severe economic crises work with large classes, poor infrastructures and low salaries.

Until recently, a lecturer in Uganda earned about US$50 per month and worked in a collapsed research culture and infrastructure. Previous governments not only deprived the university of funds; severe forms of intellectual censorship brought research to a halt. The new government recognizes the value of research and has allocated substantial sums to the university for its promotion. However, more than funds are needed as the research culture and infrastructure have become non-existent. It is proving easier to rebuild the research infrastructure than to re-establish the research culture that was once one of the liveliest in sub-Saharan Africa. As one Ugandan academic commented: 'It is so long since I did research that I almost forget how to do research. I do not know what are the recent developments in research and science education.' This notwithstanding, the same academic, together with other academics, is valiantly attempting to rebuild the research infrastructure and culture at the university.

Most interviewees felt there was a dominance of quantitative research. The reason ventured was that most senior researchers were trained overseas in the 1960s and 1970s when such work was at its height. These foreign-trained academics are now the professors who shape and control the research capacity building, direction and methodologies used. Owing to financial constraints, they lack access to international journals, cannot network with researchers from other countries, and therefore are not exposed to new research techniques. The reasons cited for doing research varied. They included requirements for a higher degree, promotion, status, financial reward and improving practice. Many beginning researchers felt the main reason for doing research was financial. They reported that donor-commissioned research was highjacked by seasoned researchers. They further claimed that commissioned research paid well and promoted a 'Pajero culture' (buying a Pajero — a racy four-wheel-drive vehicle — and other luxuries). They felt this 'Pajero culture' did not promote collaborative research. Senior academics act as gatekeepers to such funding, not as professional mentors.

One department of science education at an East African university lost three senior science educators to the AIDS virus over five years. This loss had a devastating effect on research productivity and stunted capacity building. Not only is the loss due to AIDS an economic one, it is also professional due to a loss of research expertise. One wonders what the overall effect of AIDS will be on science education research in sub-Saharan Africa.

## THE EMERGING SNAPSHOT

I use the metaphor of a snapshot to describe this study. It is limited, but presents a sketch of science education research in parts of sub-Saharan Africa that is just one possible synthesis of the data collected.

### Who and where

Mainly university academics do research. Few other persons are involved. A likely reason is that universities expect and support research as part of one's job. Most research is done as part of a higher degree or for promotion. Others in the educational system, especially policy makers and teachers, are not encouraged in the same manner as researchers.

## Working conditions

For most university academics working conditions have deteriorated. Student numbers and workloads have increased and salaries have plummeted in some countries. Ghana, Malawi, Nigeria and Uganda are such cases. In southern African countries such as Botswana, Lesotho, South Africa and Swaziland, academics work under better conditions than their colleagues elsewhere in the continent.

However, research productivity is higher in Nigeria and Ghana than in most southern African countries, despite their poorer working conditions. This suggests that it is not only conditions of work that influence research productivity. Factors such as an established research culture, a need for higher degrees or promotion, the need for experience among academics, national independence, publishing outlets, and professional associations also affect research productivity.

## Research culture

Both Ghana and Nigeria have been independent for longer than the other countries surveyed and thus have had longer to develop their research capacity. They have well-established higher-degree programmes, and both have the capacity and the experienced academics to train their own researchers. Both countries, particularly Nigeria, have active science education associations that promote both science education and research. They therefore have well-developed research cultures.

As work conditions deteriorated in Uganda — a country that once had a flourishing research culture — so did research productivity and ultimately the research culture. It would seem that, once a research culture is destroyed, productivity becomes low, regardless of the resources used to improve working conditions.

In Swaziland, conditions of work are better than in most other African countries. However, research productivity is not as high as in Ghana or Nigeria. This may be because the establishment of science education as a discipline and of research has been recent, so the research culture is in its infancy. In Zimbabwe, research productivity is lower, yet working conditions are better than in Ghana and Nigeria. Both work conditions and a research culture seem important to sustain productivity. To promote research, one must provide more than funds.

## Professional support

Most African academics do not have publishing outlets in local journals. Nigeria is the only country with a regularly published science education journal. The journal has played a central role in helping to develop and sustain the research culture. The journal is that of STAN, the professional science teachers' association. Soon after its establishment, STAN became involved in supporting its members with writing and publishing textbooks, a percentage of royalties going to the association. With a market as large as Nigeria's this has enabled STAN to become financially self-sustaining and not dependent on shrinking university and government budgets or on the changing priorities of donors.

Only one professional science education research association exists in Africa, namely SAARMSE, though many countries have national science teachers' associations. The latter may promote science education, but do little to promote research and do not act as forums for disseminating research findings. Most governments and donors do little to promote the professional development of researchers in science education.

Networking between science education researchers within Africa or with colleagues elsewhere is minimal. Networking could be promoted through meetings, e-mail, professional associations and exchange programmes. Demonstrated increases in research productivity justify investment in such forms of professional support that must be pursued more intensely.

Large research projects are undertaken only when funds are available, usually from international donors. Were they to insist that such studies are implemented collaboratively, instead of commissioning individual senior academics, they would be making a major contribution to developing research skills.

## Research skills and competencies

These vary country to country, and within any given country over time. Those institutions with demonstrated capacities have established higher-degree programmes in science education and have recognized the importance of research.

In most countries (South Africa being an exception) quantitative research is dominant. This is likely to be due to the influence of senior academics who were trained overseas in the 1960s and 1970s when such techniques were used to the exclusion of others. For financial reasons, these researchers have remained isolated, and have therefore not been able to upgrade their skills.

## What is being researched?

Research in sub-Saharan Africa tends to focus on learners, constructivism, alternative conceptions, cognition, teacher and learner attitudes, and assessment of learning, particularly at secondary school levels. A growing number of studies are focusing on curriculum reform, INSET, and curriculum instruction. Some studies, especially those appearing in international journals, focus on ethnoscience and African thought. A similar dominance of cognition, particularly constructivism, in developed countries and international journals leads one to suspect that there is a strong influence from developed countries on African science education research agendas. This may not be warranted.

Few studies focused on issues such as language of instruction, teaching in poorly resourced schools, gender, teacher education and impact research that are current priority issues in science education throughout Africa. Silences seem to centre on macro issues such as financing, equity, planning and policy. Perhaps the intellectual climate in African countries is more responsible for these gaps than the international world.

## Why is research being done, and to what effect?

The dominance of positivistic, quantitative research methodologies is an issue meriting concern, since such techniques study problems rather than explore solutions. They provide little useful feedback to policy makers, classroom teachers or educational support staff.

Most studies are short. Few take the form of baseline, longitudinal or impact research aimed at auditing the effectiveness of the system. Such a lack provides policy makers with inadequate information about what works and what doesn't. The scarcity of collaborative research impedes the implementation of large projects that could supply such information as well as train novice researchers.

As serious, according to Namuddu (1991a), is that such flawed, quantitative studies are frequently used by multinational donors to impose major change on African governments that may be neither realistic nor faithful to local concepts and perceptions. Kyle (1995) argues that, owing to the familiar split and hierarchy between researchers (the theorists) and implementers (the practitioners) it is not surprising that those on whom research is done rarely adopt research findings. Traditional research is ethnocentric, coded in inaccessible terminologies, and contemptuous of the language and realities of classrooms. Thus there is little hope of promoting change until research ideologies and practices themselves change.

The almost total absence of participatory research is cause for grave concern. African science education researchers, like their international colleagues — indeed, perhaps overinfluenced by them — engage in technical, system-maintaining studies rather than in counterhegemonic praxis. They, too, seem unaware of recent developments in postmodern and post-structuralist thought that have had a significant impact on other human sciences. Participatory research involves all stakeholders in identifying and solving problems in the education system. It demystifies research and involves more people in the process of change, including researchers. Because of stakeholder involvement, there is a much higher likelihood that findings will be implemented. Participatory research contains all three basic elements of good research referred to earlier, namely the creation, use and dissemination of knowledge, and does so in an organic way. It has an exciting potential to enable all participants in education continually to inquire into their practice and strive for improvement. A more extensive adoption of participatory research would go far in promoting the usefulness of research as a tool to improve policies and practices of science education and would provide advocacy for increased support to the research endeavour.

## CONCLUSION

The review illustrates how researchers in sub-Saharan Africa develop varying capacities to carry out research in science and technology education. These are growing fast in some countries and only beginning to emerge in others. Various factors have contributed to this uneven development. However, the impact of the research has been minimal on the improvement of policies and practices in science and technology education. Little science and technology research in sub-Saharan Africa is directed towards helping the continent face the central challenges of the 21st century.

## REFERENCES


African Academy of Sciences/American Association for the Advancement of Science (AAS/AAAS). 1992. Electronic Networking in Africa: Advancing Science and Technology for Development. Workshop on Science and Technology Communication Networks in Africa. Washington DC: AAAS

Bajah, ST. 1990. Direction of research in science, technology and mathematics education in Nigeria. In *Science Teachers' Association of Nigeria (STAN). 31st Annual Conference Proceedings*. Nigeria: Samdex Printing Works Ltd

Court, D. 1991. The intellectual context of educational research: Reflections from a donor in Africa. Paper delivered at the International Conference on Strengthening Analytical and Research Capacity in Education: Lessons from National and Donor Experience, July 1–5, Bonn, Germany

Court, D. 1983. Educational research environment in Kenya. In Shaeffer, S & Nkinyyangi, JA (eds). *Educational Research Environments in the Developing World.* Ottawa: IDRC

Court, D. 1982. The idea of social science in East Africa: An aspect of the development of higher education. In Stifel, LD, Davidson, RK & Coleman, JS (eds). *Social Sciences and Public Policy in the Developing World.* Massachusetts: Lexington Books

Evans, DR. 1994. Education policy formation in Africa: A comparative study of five countries. Technical Paper no 12. ARTS, USAID

Gibbs, W. 1995. Lost science in the Third World. *Scientific American*, August

Hallak, J & Fagerling. 1991. Educational research in developing countries: A background paper. In Strengthening Educational Research in Developing Countries. Stockholm, Paris, unpublished

Husen, T. 1990. Research perspectives: Research paradigms in Education. In Keeves, JP (ed). *Educational Research, Methodology and Measurement: An International Handbook.* Australia: Pergamon Press

International Bank of Reconstruction and Development. 1980

International Development Research Centre (IDRC). 1991. Strategic Choices for Sub-Saharan Africa. IDRC-MR289e. Ottawa: IDRC

Kahn, M & Rollnick, M. 1994. Science education research in Africa: How can it help us? In Grayson, D. Proceedings Workshop on Research in Science and Mathematics Education. Durban: UNP

Keeves, JP. 1990. The methods of educational inquiry. In Keeves, JP (ed). *Educational Research, Methodology and Measurement: An International Handbook.* Australia: Pergamon Press

Kyle, B. 1995. Research in Science and Technology Education: The Part Toward Revolutionary Futurity, unpublished

Lewin, K. 1995. Programme Support for Research on Science and Mathematics Education in South Africa: Report on a Mission 28 March–5 April 1995. Johannesburg: Foundation for Research Development; Cape Town: British Council

Namuddu, K. 1991a. Capacity Building in Educational Research and Policy Analysis: Case Study of Eastern, Central and Southern Africa. Nairobi: IDRC

Namuddu, K. 1991b. Educational Research Priorities in Sub-Saharan Africa. Strengthening Educational Research in Developing Countries. Report of a seminar held at the Royal Swedish Academy of Sciences, Stockholm, 12–14 September 1991. Paris: UNESCO and IIE

Obioma, G. 1990. New directions of research in mathematics and science education for national development. In *Science Teachers' Association of Nigeria (STAN). 31st Annual Conference Proceedings.* Nigeria: Samdex Printing Works Ltd

Reddy, V. 1995. Redress in Science and Mathematics Education in South Africa: Status of Science and Mathematics Education Research in SAARMSE. Durban: CASME

Schaeffer, S. 1983. Introduction. In Schaeffer, S & Nkinyangi, JA (eds). *Educational Research Environments in the Developing World.* Ottawa: IDRC

Sherman, MAB. 1990. The university in modern Africa. *Journal of Higher Education*, 61(4). Ohio State University Press

Tipane, M. 1990. Politics of educational research. In Keeves, JP (ed). *Educational Research, Methodology and Measurement: An International Handbook.* Australia: Pergamon Press

United Nations Economic Commission for Africa (UNECA). 1995. Development of Appropriate Science and Technology Indicators for Africa. UNESCO

UNESCO. 1994. Final Report of Symposium on Science and Technology in Africa. Regional Office for Science and Technology in Africa, Kenya, 14–19 February

Walker, JC & Evers, CW. 1990. The epistemological unity of educational research. In Keeves, JP (ed). 1990. *Educational Research, Methodology and Measurement: An International Handbook.* Australia: Pergamon Press

Yoloye, EA. 1990. Educational research priorities in Africa. In United Nations Educational, Scientific and Cultural Organization (UNESCO). National Educational Research Policies: A World Survey. Paris: UNESCO

# 12

## The mass media and science and technology education

*Tom Mschindi, Managing Editor,* Daily Nation, *Nairobi, Kenya, and*
*Sharad Shankerdass, Nairobi, Kenya*

### ABSTRACT

The mass media has a potentially important role to play in popularizing science and technology. This chapter focuses on modern mass media, traditional mass media, and their interface with informal and nonformal education in science and technology education.

### INTRODUCTION

African educators who use the media must think of traditional means of communication as well as those of the modern mass media. Traditionally, information and cultural values have been communicated by means of story telling, songs, riddles and proverbs — all oral media. Even nonverbal means of communicating such as dance, drumming and beadwork are still used. Not only are these media effective in that complex messages can be faithfully and rapidly transmitted, they are also culturally sympathetic in being participatory, more democratic and less transient than many modern media.

Mass media, literacy and urbanization are central to democratic political development. Ironically, the power of modern media as tools of mass communication is recognized by military regimes throughout Africa, since their first target on leaving the barracks is the broadcasting station. The Ayatollah Khomeini revolution was fuelled by smuggling audio tapes recorded in Paris into Iran to be played on the ubiquitous battery-powered tape recorders to audiences for whom listening to the sheik during Friday prayers was a key cultural event. No wonder that many of the fragile governments in Africa attempt tight control over media institutions, and that the masses usually distrust these top-down means of communication, putting their faith in more familiar, less manipulatable, traditional media.

Educators, including science educators, who work with mass media specialists to change behaviour in ways that empower people to take more responsibility for their own lives would do well to analyse the ways of effectively interfacing modern and traditional media. If we cannot produce material that is witty, attention grabbing and sufficiently empathetic to resonate with traditional modes of communication, we might as well give up before we start.

## CHARACTERISTICS OF TRADITIONAL AND MODERN MEDIA

All societies define and express their cosmologies. In African countries this has been done mainly through oral media such as story telling, songs, proverbs, riddles and folk theatre. Messages are also effectively coded in nonverbal media such as dance, drumming, body art and beadwork. For example:

- The talking drum of Nigeria is well known as the bush telegraph of Africa.
- The patterns of beads on a woman's apron in some Masai clans indicates she is the mother of an unnamed child. An unnamed child is not yet a clan member and such a child's hair remains unshaved. The wrath and punishment of women elders will fall on any man, including the husband, who breaks the sexual taboo that mothers of such children are under. Since only the mother can shave a child's first hair, this becomes an effective form of child spacing, placing control firmly in the woman's hands.
- The accuracy of the memories of the *griots* of Mali — the repositories and communicators of the histories of the kingdoms — is well documented.
- Before the introduction of the AK47 and M16 to war-torn Somalia, travellers were welcome at the family hearth as bearers of news, and in this pastoral society the news travelled fast.
- Shem is a dialect invented by children who roam the streets of Nairobi. It combines other languages in ways that few adults can understand. Shem has spread throughout the country in less than 10 years, and constantly invents new words. There is a similar, but not identical, dialect used by wealthier children.

The power and impact of traditional African media can be compared with that of current mass media in industrialized societies. For example:

- The impact on audiences of songs sung at community concerts in Somalia or at *tarubs* in Zanzibar, for example, compares with the impact of 'pop' songs on teenage audiences in the West. 'Rap' carries the anger of black innercity youth to white suburbia. Popular music — and in these days of electronic media 'popular' connotes billions — can even carry environmental messages.
- The earrings worn by some young men in the West can be compared with beadwork on Masai aprons. Dress, hairstyles and tattoos as identifying symbols for innercity gang members compare with markings on the shields of African warriors.
- The function of cartoon figures such as Superman, the Trudeaus or Andy Capp in defining and reflecting class values in the United States and the United Kingdom compares with tales of the heroes and heroines of African folk stories.

- The status messages communicated by the furnishings and location of offices in multinational corporations can be compared with the location of homesteads within an African chief's compound.

However, important differences must be noted between traditional and modern mass media. The oral tradition is participatory and involves a high degree of interaction between the message and the receiver. The message can be modified within accepted limits, thus making it part of the receiver's own repertoire. In turn, receivers become creators and transmitters and thus the message becomes an organic part of a people's culture. Since the originator of the message is aware of the tradition, rules must be followed for its onward transmission. Until recently in Somalia, important government memos and directives were composed as poems!

Thus traditional media contain by their nature more democratic, modifiable, and laterally transmitted messages than the centralized, top-down, impersonal transmissions used by modern media. Furthermore, the rules of traditional media ensure that the message comes across clearly. We are all familiar with modern media, where brilliant displays of style disguise a hidden content; and even when such cleverly concealed messages are detected, receivers cannot effectively express their dismay as they can when traditional media are used. Another distinguishing feature of traditional media is that, unless messages are relevant to the experiences, fears and hopes of the community, they will not be transmitted. Feedback is immediate and face to face. Of course, with modern media, one can always switch off!

Coseteng (1981), quoted in Valbuena (1986), sums up our problem as communicators wishing to use modern mass media:

> What the mass media in its [sic] high stage of development have failed to realize is that existing side by side with them on the actual village level that is quite different from the global village infrastructure . . . is another form of media, one which even antedates them — the traditional media of communication . . . Nevertheless, traditional media still survive and are used as meaningful channels of communication in traditional or developing societies. Their unobtrusive nature is, perhaps, the reason why they have been ignored for most of the time by the mass media orientated communication experts and development planners. Indeed, they are still viable forms of human communication.

Rapanoel (1991) argues that, in many societies, oral traditions are still the most important, if not the only, source and repository of traditional and popular knowledge, practices and culture. It is, therefore, futile, shortsighted, culturally arrogant — even downright dumb — for anyone to seek to facilitate change in such African communities without taking cognizance of the centrality of traditional modes of mass communication. The need to do so is more profound in this era of transition from the relatively acephalous organization of precolonial Africa to that of centrally organized nation states that are part of a world economy.

Happily, development communication planners and researchers have recently begun to restate the primacy of traditional media in mobilizing communities to make more informed judgements about using the modern technologies that increasingly impinge on their lives. A number of principles advanced by Valbuena (1986) are instructive:

'1. Folk media, modern mass media and extension services must become an integral part of any communication programme for rural development.
2. An understanding of rural audiences is vital to the effective use of modern mass media. Based on this understanding, media messages must be culturally empathetic and appealing so that communities absorb them and increase the possibility of their leading to behavioural changes.
3. There should be a synergy between traditional and modern media that leads to an emerging culture so badly needed by nation states in Africa.
4. Desired change carried by modern mass media should be sufficiently authentic to local cultures and flexible to ensure adoption.
5. There should be collaborative involvement of traditional media artists in modern media productions.'

## USING MASS MEDIA

Radio is the most pervasive medium in Africa. However, as economies decline, there have been reports that rural households are finding it increasingly difficult to buy dry cells for radios.

In Kenya, a significant percentage of the adult population is literate. The Bible and newspapers are the most commonly read publications. It is estimated that each copy of the *Daily Nation*, Kenya's largest selling newspaper is read by at least 10 people. Even so, newspapers do not penetrate deeply into rural areas. One of the first questions likely to be asked of visitors to a village is whether they have a copy of the day's newspaper.

There are mobile cinemas in Kenya. Initially an elite urban phenomenon, video playback machines are increasingly being seen in villages. Owned by wealthier households and powered by portable generators or solar panels, these become rural movie theatres, with audiences paying small fees for admission. Television remains an urban elite phenomenon.

All media in Africa tend to promulgate the position of reigning governments. Many governments have tight media laws and frequently harass journalists, close publications and smash printing presses. Often, newspapers must toe a fine line to remain open. Media oppression by African governments is frequently reported by Amnesty International and similar organizations concerned with the human rights worldwide. The popular Nigerian singer Fela Ransome-Kutie is regularly jailed, regardless of the political persuasion of the government in power. Even his Nobel prize afforded Wole Soyinka no protection when he began to write lyrics for popular musicians and agitate against the government — nor did Ngugi Wa'Thiongo's international renown offer any protection when he began to develop community theatre in Kikuyu. Audio tapes of songs with a social commentary are regularly confiscated in Kenya. Though the purpose of this paper is restricted to the educational use of media and does not address issues of political commentary, it must be pointed out that censorship of developmental messages is not infrequent. Kenya, for example, routinely bans media programmes with family planning messages.

Kenya and many other African countries have school broadcasting sections that produce programmes to supplement classroom teaching. Problems of uneven reception and maintenance of receivers aside, studies indicate they have a negligible impact on learning.

Science popularization programmes in Ethiopia did not achieve their objectives (Kebbede, 1987). Reasons included the failure of programmes to take into account local cultural beliefs and practices. Interactive radio programmes, developed by USAID as a cost-effective means of teaching mathematics, English and science in primary schools, though effective according to USAID-sponsored evaluations, were never adopted in Africa.

The Kenya Broadcasting Corporation — a state-controlled media house — has attempted to widen the scope of science-oriented programmes through programmes such as 'Panorama', 'Science Digest', and 'Tomorrow' on radio and television channels. Even these local productions are elitist and are of little interest to the urban poor, farmers and the struggling middle class, who need all the information they can get to improve their economic and social conditions.

Some government departments, such as Ghana's ministries of health and agriculture, produce simple, well-designed and relevant pamphlets for villagers. However, budgets are such that they reach an insignificant portion of their target audiences. Newspapers penetrate more deeply and consistently, and frequently have regular columns devoted to science, health, agriculture and appropriate technology. Unfortunately, these articles are too often written as if the reader already has a sophisticated knowledge base and they consequently read more like research publications. This approach assumes that what is needed is more information from which readers can select whatever is applicable. This in turn can become an argument for such coverage to be given more space!

Scholars such as Metere (1991) posit that science and environmental issues will be more professionally, intelligently and sympathetically handled only if journalists are recruited to the cause and specifically trained. In the 1970s and 1980s, the International Development Research Council (IDRC) supported a series of regional workshops to train science writers and editors. Though the programme may have contributed to improving the quality of scientific journals, it has had no visible impact on writing in the popular press.

## THE WAY FORWARD

Africa is in transition, and for those of us concerned with education and development, the mass media is a messy business. This messiness opens up new possibilities.

The earlier discussion emphasized the need for a fusion between traditional and modern media. Other factors emerge as one continues to analyse communications issues in Africa, such as differences in producing material for an information-rich society and for one that is information-poor.

Information-rich societies store information in museums, libraries, electronic data bases, and so on. By definition, newspapers and many other print products are ephemeral. In an information-poor society, the human mind is the community's data base. However marvellous the ability of elders and *griots* to memorize knowledge, they cannot compete with what was made possible by the invention of the Gutenberg press and the microchip. In information-poor societies, newspapers and magazines are given brown papercovers, become dog-eared with use and are kept for decades.

In industrialized societies, time is at a premium. People always seem to be in a hurry to get somewhere. Billboard messages are designed to be read by people in vehicles rushing by at speed. Radio and television prime time may cost hundreds of thousands of dollars per minute. By contrast, African film audiences are outraged by advertisements that do not take time to tell stories. (Ironically, at least in the United Kingdom, some highly-talked about television advertisements have begun to tell elaborate stories over successive broadcasts!)

Such analyses can help us to invent ways of using modern media more effectively in Africa. I suggest the following categories of mass media materials to promote scientific and technological understanding:

- *Motivational materials*. These should encourage thought and debate on science, its role and utility. They should help people understand why they should consider learning more.
- *Substantive knowledge*. These should contain information ranging from the uses of the neem tree (*Azadirachta indica*) to instructions on how to repair a car engine.
- *Methodological materials*. These should motivate and encourage teachers to use inquiry approaches to teaching.

## Financing

Unlike industrialized countries, Africa has a scarcity of funds for educational media production. In the past, the state has funded most radio and television productions and budgets are slender. The private sector subsidizes programmes through advertising agencies that have little experience in using media in innovative ways. Larger private-sector firms are just beginning to insist that agencies work with smaller companies or NGOs that have demonstrated track records as innovative communicators. For example, a large multinational with its regional headquarters in Nairobi has recently insisted that its advertising agency subcontract a significant proportion of its account to a small, innovative publishing house. A campaign will target rural communities through schoolchildren: it will hold competitions nationwide and provide supplementary educational material to winning schools and students. Schools, students and parents will benefit; the multinational will improve its public corporate image, will demonstrably be seen to contribute to nation building, and its name will be carried to all households with school-going children in the region.

A curriculum development project, Science and Technology in Action in Ghana (STAG), based at the University of Cape Coast, has attracted substantial funding from large industrial and manufacturing companies. It did so, not by directly appealing for funds but by being asked to contribute professionally towards developing curriculum materials for upper secondary schools. When senior industrialists came to the university for writing workshops and experienced the working constraints directly, they competed with one another to pay for desktop publishing equipment, the printing of project materials, and so on. A similar project in Swaziland is also gaining attention and support from the private sector.

We must continually search for similar innovative ways to fund media productions.

## The target audience

All too often we are trapped into preconceived notions about what constitutes the 'mass' audience. In the past, the image of Africans has been that of the generic rural 'peasant'. But rural communities are changing fast. For example, in Western Kenya the primary health care programme is mostly being run by retired civil servants. This group is a rapidly growing target audience that can be used by modern media as an entry point to traditional modes of communication. In the 1990s, there are few households where no-one can read. We must be sensitive to the changing circumstances of our communities as they stratify and become less homogeneous.

## Production strategies: print

Whether we are media specialists, scientists or curriculum developers, our challenge is to design multimedia strategies that penetrate the market, are empathetic and can be absorbed by traditional modes of communication. It is worth analysing some examples.

### *'The Young Nation'*

The *Sunday Nation* has the largest circulation of any newspaper in Kenya. On Easter Sunday 1994 it began to carry a supplement targeted at the youth, 'The Young Nation'. Initially it was planned as a double-page insert to be published once a month. On the day of the first issue, for the first time in its history, the *Sunday Nation* sold every copy printed. The next morning, the *Nation*'s switchboard was inundated with telephone calls from the private sector wanting to place advertisements. One long-distance call was from the paper's proprietor, His Highness the Aga Khan. This, his first telephone call in years, was to congratulate the paper's managing editor on the 'The Young Nation'. The same morning, *Nation* staff contacted the innovative children's publisher to whom they had subcontracted design of 'The Young Nation' to ask if they could produce a larger supplement to be published weekly. Since then

'The Young Nation' has become a 12-page spread, published weekly. Increased sales more than justify its publication.

### *The* Kenya Times

For about a year the circulation figures of the *Times* — the daily newspaper of the ruling party — were significantly boosted by a weekly mathematics supplement targeted at children in primary school. Sadly, changes in senior management led to its demise.

Supplements could be used more extensively to satisfy needs in information-poor Africa. Few rural households in Victorian Britain and the United States were without their homesteader's almanac that contained information on almost everything they might need. African newspapers could carry supplements containing relevant information and household tips on subjects such as health, agriculture and appropriate technology. Major European novels by authors such as Dickens were often first published in broadsheets, chapter by chapter; similarly, newspaper supplements could supply rural Africa with leisure reading, encourage promising authors, and simultaneously promote sales.

### *The Kagera Writers' and Publishers' Cooperative Society (KWPCS)*

This group in a remote region of Tanzania publishes a monthly newspaper that carries development messages. It intends to develop a supplement for youth similar to 'The Young Nation'. Stringers in over 70 villages are paid for published copy. To distribute the materials — notoriously difficult in Africa — the cooperative 'piggy-backs' on farmers' cooperatives that have the necessary infrastructures such as warehouses, lorries and centres at a density of about one to every three primary schools. The KWPCS uses the same distribution system to sell the supplementary materials it writes and publishes for schoolchildren.

### *The South Africa Newspaper Education Trust*

This NGO prepares copy for a weekly supplement carefully designed to help teachers. Placement of the supplement in a nationally distributed newspaper guarantees wide distribution. Initially donor funds are being used to help develop copy, to train writers, and to ensure that multiple copies are sent to schools in deprived communities. Detailed business plans have been drawn up to enable it to become financially self-sustaining.

### Action Magazine

*Action Magazine* is an innovative solution to the problem of reaching the resource-poor classrooms of Africa. It has created an institutional framework in which staff from curriculum development centres in Zimbabwe, Botswana and Zambia work together with graphic artists from *Action Magazine* to produce a health/environmental magazine. *Action Magazine* is designed to appeal to children, using graphics, cartoons, stories, games and competitions. Topics are selected around the syllabus.

*Action Magazine* has set up a direct mailing system to all primary schools, secondary schools and teacher-training institutions in Zimbabwe and neighbouring countries.

### *Mazingira Institute*

The Mazingira Institute uses a similar strategy of mailing supplementary materials directly to schools in Kenya. It has set up an effective system of disseminating material on topics such as immunization, marine pollution and conservation. Mazingira designs four side supplements in an amusing, accessible manner. Once the artwork is ready, Mazingira buys space in leading newspapers such as *Nation*. Over a million readers are invited to take the supplement for their own use, or to pass it on to teachers they know. It is known that messengers and secretaries in cities keep the supplement and take it back to their homes in rural areas to give to children and local schools.

### *The Somali Family Health Project*

Before the disintegration the infrastructure in this war-torn country, traditional and modern media of all sorts were effectively used to promote family health. Particularly interesting was the way the project used print to tap into traditional communications networks in this society where poets — male and female — were more highly valued than warriors. Single-page leaflets with cartoon stories carried development messages. Heavily illustrated novellas carried messages to adolescents and youth. Posters depicting culturally familiar icons such as a nomadic family on the move, reminded people of their cultural and religious traditions, while publicizing social issues, such as the importance of child spacing. Posters and T-shirt designs exploited the Somalis' love of debate. Elaborate and strange designs were employed. One such design for women's attire depicted on the front, a mother, protectively clasping her baby who was threatened by a six-headed snake — each head showing symptoms of an immunizable disease. The back showed all six heads pinned to the ground by the syringe-shaped spear carried by an angel of mercy dressed as a health worker. Development communications experts thought this design much too sophisticated until they saw what a crowd-stopper it was on streets and in markets. In tea bars, adults could be seen playing board games involving health messages the same evening they had been supplied to primary schools in the area.

These examples show that new desktop publishing systems can rapidly provide innovative, culturally sympathetic science and technology materials. The NGOs and private-sector organizations concerned showed how curriculum developers can be involved, how money can be raised and how new institutional linkages can be created to overcome some constraints on the production of printed material in Africa.

## Production strategies: video

### *'Spider's Place'*

Production costs for television and video can be underwritten by aid agencies or the private sector. A good example is 'Spider's Place', designed by the Handspring Puppet Company, a South African NGO. Spider, a strong-willed young woman, is the leader of a gang of puppet characters whose ingenuity in science and technology gets them out of innumerable scrapes. The programmes are broadcast and disseminated on VHS to areas not reached by television. Comics and radio broadcast adaptations reach areas where video playback facilities are not available.

### *Linking School with Community Science*

Linking School with Community Science is a project implemented by the Malawi Institute of Education together with training college tutors and key Ministry of Education officials. It is a multimedia project designed for both in-service and pre-service teacher development. The first video in the series shows village craftspersons such as the brewer, potter and thatcher being sensitively interviewed about their understanding of the science and technology underpinning their trades. Other tapes in the series show teachers working with primary schoolchildren using this community knowledge. Appropriate booklets accompany each video.

It should be noted that most African countries have well-equipped but moribund educational media centres. These were established in the 1970s. Many have insufficient funds to continue work, yet there is no reason why new ways of funding should not be encouraged. NGOs and donor agencies could be asked to support video projects along the lines suggested by the work done by *Action Magazine*, Mazingira and Handspring.

Broadcast-quality productions are prohibitively expensive. But with technologies such as Hi8 and VHS there is no reason why production costs cannot be lowered. It is now possible for teacher training colleges to produce methodology videos. An example is the work described in Malawi. In Zanzibar, such equipment is used by some cluster teacher groups to promote inquiry and gender-sensitive teaching by recording and critiquing one another's classrooms. It should be noted that many broadcast stations in Africa now collect their footage on Hi8. This means that coproductions can easily be organized.

In coordinating with media specialists, it is important that curriculum developers are not overimpressed by their demands. Media producers often favour sublimely beautiful images over content. The exaggerated images they incline towards are often terrifying to children. All that is required for educators to learn to make their own programmes is a basic understanding of video technology, and the grammar of film.

There is no point in making videos unless they can be viewed, however. The VHS revolution makes it easier to show videos to communities. Video parlours can be hired or borrowed. But it is critical that the dissemination strategy is designed before shooting begins. There is also no point in making anything the audience cannot understand or associate with. This means that, however correct the content, the form must match the cultural aesthetics of the viewer.

Discovering the appropriate aesthetic framework can be hard work. For example, after many years of making movies for rural communities, it suddenly dawned on one of the authors that the target audience did not appreciate abstracted, disembodied commentaries. Further reflection made him realize that most village communication was oral. People listened and talked to each other, or sang songs. Thus he made a programme with no commentary. All substantive information was imparted by people on screen talking to one another, being interviewed or through songs. When these videos were shown to village audiences, they understood perfectly.

### Production strategies: audio

Song features strongly in many African societies. A brilliant primary health care project in Western Kenya uses live performances or audio tapes to carry messages in the form of songs in the local language. It is interesting that the community readily relates to the songs, frequently breaking into dance, yet they always remember the messages.

A project of the Malawi Wildlife Services uses theatre organized by village youth to stimulate village discussion about better use of local resources. One competition organized in a district township drew a crowd of over seven thousand!

Africa has yet to see the 'soap opera' being used to promote development. 'The Archers', a well-known radio broadcast in the UK, was started shortly after World War II specifically to carry culturally sympathetic and appropriate messages to farmers. Similar radio programmes in South America, initially supported by funds from donor agencies, devoted over 40 episodes to establishing characters and plots before any hint of development issues was introduced. 'The East Enders', a British television soap opera that has been broadcast for decades, was never intended to carry social messages. Eventually its producers realized its power to do so, however, and issues such as AIDS are now interwoven in the ongoing plot. These broadcasts lead to measurable changes in the behaviour of viewers. Radio programmes such as these are not expensive to produce, but have not yet been exploited in Africa.

## CONCLUSION

In discussing the need to publicize the nature of science to the peoples of Africa, it is important to point out that scientists and curriculum developers can be their own worst enemy. Not only is their writing opaque; most of them never really try to reach a wider audience. Yet, with a little energy, they could garner publicity for their work. At the least, they could approach media specialists to cooperate on joint productions. Doing so would (1) help to inform the general public; (2) motivate the public to appreciate that science and technology are relevant in African society; and (3) gain sympathy for science and scientists who are everywhere under attack.

All too often — whether we are scientists, curriculum developers, or indeed media specialists — we treat the production of educational media products too lightly. The same amount of care is needed in such productions as goes into the production of research papers, curriculum materials or major pieces of investigative journalism. Perhaps no one put this better than Jean Luc Godard when he said, 'If films were airplanes, most people would die!'

## REFERENCES

Coseteng, ML. 1981. Traditional media in developing societies. In Valbuena, VT (ed). 1986. *Philippine Folk Media in Development Communication.* Singapore: Parklane Press

Kebbede, T. 1987. Popularization of Science Technology in Ethiopia. Current Situation and Future Directions. Paper presented at the Science Popularization Research and Services Council of the Ethiopia Science and Technology Commission, Addis Ababa

Metere, A. 1991. Health and environment concerns in Africa. In ACCE. *Module on Specialised Reporting.* Nairobi, Kenya: ACCE

Rapanoel, D. 1991. The contribution of oral traditions and mother tongues to the communication strategy in rural communities. In Brajo, K & Geogr, N (eds). *Communication Processing. Alternative Channels and Strategies for Development Support.* Ottawa: IDRC Communications Division

Valbuena, VT (ed). 1986. *Philippine Folk Media in Development Communication.* Singapore: Parklane Press

# 13

## Into the next millennium

### INTRODUCTION

This chapter attempts to synthesize the preceding chapters and summarize discussions at the ASTE '95 meeting, not only those in formal sessions, but also those that raged deep into the night. The chapter cannot do so faithfully. It is not that we do not represent what was discussed — on occasion debate was heated and many opinions were expressed — but we have our own prejudiced ears. We hope that our biases do not come through too strongly. However, assisted by scribes who took notes throughout the meeting, we hope we have captured its spirit. The synthesis focuses on the challenges and the way forward for science and technology education in Africa for the next millennium.

### SCIENCE, TECHNOLOGY AND DEVELOPMENT: PRE-EUROPEAN CONTACT

#### Technology

Technology is intimately interlinked with the development and evolution of our species. Our first home was Africa and only with the Industrial Revolution in Europe did our family paths part so dramatically. Today, when the ways and products of science and technology dominate world culture, Africa, our cradle, has been left a victim, spectator and consumer.

There is no doubt that technology was a central element of African cultures (Makhurane and Kahn, chapter 2). The pyramids and iron mines of Meroe in the north, bronze sculpture and earthworks of the rain-forest kingdoms of the west, and the ruins of Greater Zimbabwe bear witness. Steel produced by the Bessemer furnaces in Europe was not necessarily better than that produced by the huge, bellows-driven, clay furnaces in Tanzania — but it was a great deal cheaper. The practice in Somalia of scratching people with pus from cattle infected with cowpox was effective against smallpox — but industrially produced vaccines were more consistent. Rice farmers in Sierra Leone carefully selected rice with hairy husks, despite their smaller size, to protect their crop from birds — even though, like farmers everywhere, they knew nothing of genetics.

## Science

The extent to which science flourished in an Africa that lacked the Gutenberg press to make it a public rather than a private practice is also debated. Before the invention of the Gutenberg press, even in Europe, science was engaged in by the few, and superstition rather than rationality was the dominant mode of thought — as is the case in much of Africa today.

Some social constructivists (Jegede, chapter 10) argue that all knowledge, including science, is socially constructed, must be defined within specific cultural contexts, and is not therefore universal. They claim there is a Japanese science, an African science, an Indian science and so on. Others, including Makhurane (chapter 2), align themselves 'with a rationalist view of science as a culture that may be superimposed on any culture since it is universal, and a culture of hope and undying optimism'. Some suggest that science developed in the north through sponsorship by the emerging elite and middle class that arose from, among other factors, land enclosure acts. In most of Africa subsistence farming even today does not lead to such class formation, and therefore Africa lacks the leisure and sponsorship needed for the growth of science.

Universal science may now be associated with the industrialized north, but historically it built on and co-opted science from areas such as Egypt, the Middle East, India, and China, and those regions into which Islam spread. Indeed, even today, Western or universal science continues everywhere to pay close attention to traditional practitioners such as herbalists. 'However, history dictates that societies keep at the cutting edge of technology to avoid domination by others. One may be first through the technological door, but that advantage must be nurtured and maintained. Those who know iron are likely to dominate those who know flint' (Makhurane and Kahn, chapter 3). Or, as put succinctly by the French poet and gunrunner in Ethiopia, Verlaine, 'Whatever happens we have got the maxim gun and they have not'.

Science as practised traditionally in Africa, or indeed in places such as India, China and Polynesia, tended to be anthropomorphic and ecologically based. Science enabled people to live in harmony with the natural environment and they saw little need to advance their science. Yet in India and China today, universal science is practised extensively — integrating knowledge from traditional science when this proves useful, such as the sophisticated system of detecting earthquakes in China. In Africa, metaphorically we too must leave the age of flint for that of iron: we need the maxim gun to avoid marginalization.

## SCIENCE, TECHNOLOGY AND DEVELOPMENT: POST-EUROPEAN CONTACT

### The realities

Certainly, since independence, African leaders have repeatedly stated their hopes for the contribution that science and technology can make to development in a continent struggling with an unfriendly environment and the legacy of slavery, imperialism and European mercantilism. These expectations have not been fulfilled. Instead

GNPs have fallen steadily. Cheru (1989) states, 'In 1981, according to the World Bank, 29 of the 36 countries with the lowest GNPs were in Africa. Seventy out of every 100 Africans are either destitute or on the verge of poverty ... One out of four Africans has access to clean water. Of the 33 million people added to the workforce during the 1970s, only 15 million found remunerative employment.' Drought, civil wars and poor governance play their part. So do policies determined by the north favouring the production of cash crops to earn foreign exchange, and import substitution manufacturing to save it. A dramatic drop in prices for primary products and a rise in costs of imports have reduced Africa's capacity to feed itself. More funds currently leave the continent each year in debt repayments than come in through technical assistance and investment.

## The needs

### *High science*

The continent needs more minds capable of engaging in high science, such as those at the International Institute for Tropical Agriculture (IITA) in Ibadan, Nigeria, where breakthroughs were made with root crops; at the International Centre for Insect Physiology and Entomology (ICIPE), Kenya, where biological ways to reduce insect damage to crops and livestock were developed; and those scientists in South Africa who have made basic contributions to cosmology.

### *Applied science*

The continent needs people capable of engaging creatively in applied science, such as those at the University of Kumasi, Ghana, who set up consultancy services in the heart of industrial areas and sell their services to groups of village women and major industries equally effectively, or those scientists in South Africa who have produced petroleum from coal.

### *Low science or science for all*

The continent needs more entrepreneurs capable of engaging in low science, such as those in the 'Jua Kali' sector in Kenya who improvize in most ingenious ways. Above all, the continent needs citizens able to make more rational choices about the utilization of their resources and the technologies that increasingly impinge on their lives.

## Nurturing science and technology

There was debate at the meeting of how science and technology could best be nurtured in sub-Saharan African countries to bring them back to the cutting edge, whether it be by drawing on African science, on universal science or on both. That there is a need to do so was unanimously agreed. We think there is a need for 'high science' in Africa. The problem is one of resourcing and of choosing fields where Africa has pressing needs or a comparative advantage. For example, the physics faculty of the University of Cape Town is known worldwide for its work in cosmology, a field that requires few resources. Hardly a pressing need, but as a rigorous training ground for generations of inquirers and future scientists, some of whom will engage in work that does address needs, the faculty has proven its worth. Breakthrough work at ICIPE and at the Consultancy Group in International Agricultural Research (CGIAR) centres located in Africa requires high levels of funding. These centres do more than practise 'high science'; they provide a haven for some of the best African scientists and training for postdoctoral students, through links with national institutions do much to keep 'high science' alive in the continent, and, equally importantly, such centres validate the practice of 'low science'. These centres are expensive and are currently funded by donors. Perhaps a day will come when African governments see the value in pooling resources to support similar regional scientific institutions.

'High' in the term 'high science' is simply a way of indicating how high up on the library stack the knowledge is; to access the knowledge and make creative use of it requires some familiarity with work at the cutting edge. Participants at ASTE '95 thought that with appropriate training those working in 'low science' on development problems could access 'high science' through programmes such as sabbaticals, exchange fellowships and jointly implemented research. But appropriate policies and resources must be available.

## WHAT SCIENCE AND TECHNOLOGY EDUCATION?

The failure of education to root scientific and technological thinking in the African consciousness is responsible for the state of science and technology practice throughout much of the continent. In his analysis of science education efforts in Africa, Yoloye (chapter 1) suggests that we critically review the legacy of past efforts as a basis for future action. Yoloye identifies vision and human development — developed either through working with mentors or in specialized institutional programmes — as elements that endure. He suggests we stop thinking in terms of success and failure and of expectations for rapid change, and instead build slowly on pockets of good practice.

### Inquiry-based science learning

What do we mean by success and failure, and by good practice? Few of us, unlike the World Bank, would use as indicators performance rates on memory-oriented examinations we all deplore. It is difficult to describe good practice. Yet when we see children doing the best with their minds, no holds barred (to paraphrase Nobel Laureate Percy Bridgeman), we all agree they are learning good science. One of us, Savage, described examples of good practice in detail (chapter 3) to provide a common starting ground to convince participants at ASTE '95 that there is nothing inherent in teachers, or students, or cultural expectations to prevent inquiry learning in classrooms in Africa. Evening presentations by participants of ongoing work confirmed that inquiry learning is possible in the environments of their projects. Most participants agreed that inquiry science was better training than rote memory teaching for 'high-science', 'applied science', 'low science', plain old tinkering and problem solving back on the farm, or for employment in the formal and informal sectors. Inquiry permits learners to bring their knowledge, language and culture to the classroom, thus moving from parallel to secured collateral learning as urged by Jegede (chapter 10), who considers learners' world-views as strengths upon which to build rather than as handicaps. Inquiry promotes relevance since investigation must be of available materials and issues (Rollnick, chapter 5). Inquiry moves authority from textbooks and teachers to pupils' abilities to marshall evidence, thus facilitating equity (Reddy, chapter 6). Inquiry learning promotes a critical scepticism, respect for others' opinions, and the ability to work cooperatively (unlike the competitiveness encouraged by traditional teaching), thereby promoting good citizenship.

### *Inquiry learning and the promotion of relevance*

Relevance was heatedly debated. Relevance affects the quality of science education, yet defining it seems fraught with difficulties, since relevance is a function of time and socioeconomic realities. There was a time in Africa when qualifications, regardless of their content, made the difference between having the dust thrown in one's face by the car, or being the driver. Even today, parents value the qualification rather than the education. By and large so do the educational system and employers. The middle class in Kenya view the introduction of the 8–4–4 system and prevocational skills as a waste of time, since their children will not need the skills, and consider the change as a reversion to the colonial curriculum. Rural families view the changes as a waste of time, since village economies already have such expertise, and as a move to block their children's access to the middle class. All view equipping the schools as an additional financial burden.

Relevance is determined by culture. Culture changes with socioeconomic or political change. Thus notions of relevance change, which is a sound reason for science educators in African countries to view solutions developed elsewhere with caution. However, some elements of relevance to universal science remain constant, as they pertain to the discipline. Thus, recognizing that specialization would be necessary at upper secondary and university levels, participants generally supported a notion throughout primary and lower secondary school of 'science and technology for all' that stems from children's interests, promotes inquiry into their natural and social environments and uncovers connecting scientific ideas. Participants were unanimous that appropriate examinations and teacher development programmes must accompany such courses.

### *Inquiry learning and the promotion of equity*

Relevance is related to equity since measures to promote equity advocate making opportunities available to disadvantaged groups — people deprived of opportunities or discriminated against on the grounds of gender, race, class, or their rural/urban circumstances — to obtain an education that is assumed to be relevant to a dominant group. South Africa is a striking example of this deprivation theory. But in a sense, any education serves the elite, since all aspects of society support its continuing dominance. Even the new South Africa cannot support a society consisting solely of an elite and though compensatory programmes may be necessary in the short term, they are an impossible long-term solution. Providing equal opportunities does not ensure equity. Equity means that everyone must be equally prepared to take advantage of the opportunities, and doing so is a societal problem beyond the control of educators. Participants thought 'first get science education right, then implement equity measures if necessary'. They generally agreed that inquiry science promoted both relevance and equity at the micro-level in classrooms (Volmink, chapter 4).

Disagreement came when participants discussed whether inquiry science is possible within the constraints of African classrooms today and, if so, how to change practice. Inquiry learning requires an enabling environment that African governments increasingly find hard to provide, and which the World Bank argues is unnecessary since it does not affect performance in examinations (Onwu, chapter 8).

## HOW DO WE CHANGE?

The meeting identified large classes, few resources, poor teacher education and centralized examinations as limiting what is possible in classrooms throughout Africa. Onwu (chapter 8) addresses large classes, Fabiano (chapter 9) looks at what can be done with limited resources, and Dyasi and Worth discuss teacher education (chapter 7). The meeting also identified a lack of suitable learning materials and examinations that stress memorization as other important limiting factors. Participants agreed that all change must be supported by major policy change.

### Who controls the discourse?

Volmink (chapter 4) examines who initiates and influences educational change. In his own country, South Africa, previous governments used power to entrench white elites and to transfer power from one group of whites to another — arguably, the promotion of Afrikaners was the most effective use of affirmative action the world has seen. The recently elected democratic government's immediate concern is again to redress the imbalance. Similar political issues have been the basis for educational change in other African countries. Immediately after independence, the role of education was to replace the class of colonial professionals and managers; later expansion was often to redress tribal inequalities. And there are always groups that cry, 'Foul, standards are falling!', as they lose control of the discourse.

Yet policy can change the playing field (who needs to make it level?). In the United States, for example, decentralization leaves poorer communities without the resources to help their students meet national standards set by the same people who have moved to the suburbs. In Africa, private education has become a growth industry for the middle classes, who have opted out of the discourse on public education. To the extent that academics in Africa are involved in the discourse, they work within stated or unstated parameters set by the government in power. Or, as both Naidoo and Rollnick (chapters 11 and 5 respectively) suggest, since many African scholars receive their training outside Africa, publish in Western journals, and continue to view this community, rather than Africa, as their reference group, their role in the discourse often addresses the international academy rather than the realities of their own situations. As we go to press, 'Teachers sidelined in new curriculum planning', reads a headline in the *Mail and Guardian* of 24 December 1996, one of South Africa's most respected newspapers. The article argues that the curriculum is being developed by 'outside specialists, ... mainly driven by the labour and training sector'.

In other countries of sub-Saharan Africa, deteriorating economies and increasing reliance on the World Bank have led to the bank's economists and cost-effectiveness experts influencing decisions increasingly and directly through the introduction of technical solutions to educational problems, and indirectly through structural adjustment programmes (Rollnick, chapter 5). As a result, teachers are rapidly becoming marginalized in favour of the textbook, and in any case their salaries are so low that they have to seek additional employment — often, ironically, doing private coaching with facilities denied them in their regular classrooms. Volmink argues for wider participation in the policy discourse through decentralization — and only policy can permit this. Until syllabuses are written that stress scientific inquiry skills and broad concepts that can be developed using materials in local environments, teachers, parents and even students can never participate in decisions about practice.

## Few resources

Fabiano (chapter 9) eloquently argues for more funds for science and technology education, but acknowledges that their allocation is unlikely, and that existing resources must be used more effectively. Though recognizing the importance of pre-service education, he presents a case for the cost effectiveness of appropriate school-based professional development that enables strained economies to control the rate of growth of qualified teachers; a decentralized curriculum that permits use of local materials; provision of equipment that can be used for 'minds on' activity rather than to confirm theory and that can be used in many ways rather than for one-time demonstration. Fabiano further advocates more use of thought experiments, and equipment and learning materials that promote inquiry; and proposes incremental rather than radical change on the ground of cost effectiveness. All participants at ASTE '95 argued for more innovative and effective ways to maximize the use of existing resources.

## More effective ways to teach large classes

Onwu (chapter 8) recognizes the reality of large classes in Africa. Despite high percentages of national budgets being allocated to education, increasing enrolments have led to a decreasing allocation per pupil. The problem is exacerbated by influential World Bank correlation studies which show that large classes having little or no effect on pupil achievement in rote-memory-oriented examinations. As Onwu points out, in such situations it matters little whether classes contain five or 50 pupils since rote learning from textbooks and notes written on chalkboards by school monitors are effective ways to teach facts. He suggests that further work is required on teachers' own classroom research and coping strategies, as well as action research that develops more effective ways to teach large classes. All participants agreed that there is a need for policies and structures, appropriate materials and examinations, that provide teachers with a better enabling environment. And all participants agreed that more effective teacher education is the basic issue.

## Teacher education

Dyasi and Worth (chapter 7) strongly propose that inquiry should be the basis of science learning. The professional development of teachers, they argue, should be ongoing. Teachers must themselves be exposed to inquiry into phenomena, their own learning, and to facilitating such learning in classrooms as well as in strategies of change.

## The mass media

Mschindi and Shankerdass (chapter 12) challenge the media to contribute to the promotion of science and technology. They claim that modern media must be culturally sympathetic and resonate with traditional modes of mass communication such as dance, song and story telling. Africa is information poor, unlike industrial countries where ephemeral print, television, radio and electronic messages constantly bombard households, and requires a different use of media. Distribution is as important as production. Mschindi and Shankerdass suggest the use of newspaper supplements to help householders build reference material similar to the almanacs referred to by Western families in the Victorian era, and as resource material for teachers. They advocate appropriate television and radio soap operas, travelling theatre, and wall art. They applaud a recent interest by the private sector in sponsoring educational media productions.

## The role of research

Little will be possible without research, and the participants at the meeting defined research broadly. Naidoo (chapter 11) reviews African research capabilities, urging that they challenge and critique policies and practices emanating from the donor community, and focus on those that address African realities. He regards questioning the purpose of schooling, the role of research, and ensuring scientific literacy for all and harmony between school reform and that of teacher education as priorities. To overcome distinctions between theory and practice, Naidoo argues, a need exists in Africa for more participatory and collaborative action research to change the lived experience of teachers and learners. Though facilities in research institutions have deteriorated throughout Africa, Naidoo identifies the lack of a research culture as being as important a limiting factor as the lack of funds. Participants argued that to restore the research culture throughout the continent there was a need for sponsorship of group research, and fellowship programmes in institutions that have a lively research culture.

## INTO THE FUTURE: RECOMMENDATIONS

Recognition that all work must be firmly contextualized within the African environment, involve the major stakeholders, and that change is an incremental and lengthy process underlies all the meetings, discussions and recommendations. Participants expressed an opinion that, despite problems facing science education in Africa, there was cause for cautious optimism. During better times Africa experienced much of what the meeting recommended (Yoloye, chapter 1 and Savage, chapter 3). We know that the implementation of inquiry science is possible. It is to be hoped that times will change again. Meanwhile, there is throughout the continent a resurgence of innovative practice. The inability of the state to maintain education itself is reason for cautious optimism. In the vacuum, others are increasingly entering the arena: NGOs, the media, the private sector and schools themselves are beginning to evolve innovative ways to support education (Savage, chapter 3). Based on deliberation throughout the week, participants made the recommendations that follow below.

### A regional centre

Participants recognized that recommendations require systematic and persistent follow-up that can most effectively be implemented by a regional science education organization such as the African Forum for Children's Literacy in Science and Technology (AFCLIST). AFCLIST has demonstrated its viability by the impact it has made on thinking and practice in African science education. ASTE '95 urged professionals, policy makers and donors to provide the support necessary to enable AFCLIST to continue to play the supportive and catalytic role necessary for science educators in Africa.

Currently an activity of the Rockefeller Foundation, AFCLIST plans to register as an independent body with secretariats based at the University of Durban-Westville and Chancellor College, the University of Malawi. To guide policy, AFCLIST has an advisory board consisting of experienced scientists, educators and media personnel. A grants committee recommends proposals for funding. Through its small grants programme, AFCLIST has supported over 60 projects in 11 countries including Sierra Leone and South Africa, in both formal and nonformal settings. In addition, AFCLIST supports networking activities such as a newsletter, interproject visits, skills workshops and meetings such as ASTE '95.

Participants at ASTE '95 recommended that donors support nodes or centres of excellent practice to be associated with AFCLIST to ensure capacity building throughout Africa.

## Resourcing science education

A creative use of limited resources is critical in educational institutions throughout Africa. Participants recommended support for projects that:

- Decentralize elements of the curriculum to encourage more use of local materials and expertise.
- Develop supplementary material in newspapers for teachers and students.
- Involve the private sector in the provision of educational materials.
- Encourage school-based teacher development groups.
- Develop training colleges as centres for resource management.

## Nodes or centres of excellent practice

### *Curriculum innovation*

Participants questioned the viability of inquiry learning within the context of the deterioration experienced by institutions throughout Africa today; however, they applauded the vision. The establishment of a node for curriculum innovation would catalyse work that would include the promotion of:

- More flexible institutions for curriculum change such as NGOs, the private sector, teachers' associations and temporary alliances for change that should involve all stakeholders.
- A critical review of past and current innovations to better guide future practice.
- Networking of good practice through meetings, newsletters and journals, electronic media and so on.
- Action research that develops models of good practice, including innovative use of media.

### *Policy research*

Policy research should focus on improving practice. Participants recommended support for a node for policy research that:

- Identifies, develops and disseminates exemplars of good practice at primary-school, secondary school and teacher education levels.
- Analyses impediments to good practice and suggests effective alternatives.
- Informs policy makers and managers of ways to facilitate good practice.

### *Teacher education*

Considering the central role of teachers, participants recommended support for a node for:

- Documentation and dissemination of exemplary practice.
- School-based teacher development projects.
- Innovative pre-service and in-service teacher education projects.
- A regional project to develop approaches and model materials for pre-service and in-service teacher development.

### *Examinations*

Assessment should reflect the spirit and goals of the science curriculum and encourage best classroom practice. The reality is that examinations are a major constraint on the development of good practice. Participants recommended that governments and donors support a node for:

- The collection of relevant data, literature, workshop proceedings and exemplars within Africa and elsewhere for dissemination to appropriate target audiences, such as policy makers, examination staff and curriculum developers.
- A regional item-writing and examinations-construction workshop for small country teams of appropriate staff to be followed by in-country research and workshops.

### *The media*

The media should be more extensively used to promote good classroom practice, support teachers and provide relevant information. Participants identified a need for models that use both traditional and modern media.

### *Teaching large classes*

Support should be given to a node that facilitates work which suggests ways for teachers of large classes with limited resources to better promote inquiry learning.

### *Gender studies*

A node should be established to:

- Monitor gender equity within AFCLIST and in the science and technology system in Africa.
- Develop and execute gender sensitization training.
- Identify and set priorities and an agenda for research and development.
- Develop, execute and research demonstrative gender interventions.
- Undertake capacity building of researchers and activists, and encourage the development of more centres for gender equity on the continent.
- Establish a resource centre for gender equity that promotes networking and dissemination.
- Mobilize resources (financial, human and physical) to sustain the centre.

The analysis of recommendations is neither definitive nor exhaustive. In this book, for instance, we have not focused on information technology and its potential impact on development in Africa.

Most developed countries of the North have followed development paths from agrarian to mining, to industrial and manufacturing phases, and are entering the information technology phase. Most African countries, however, have not yet entered the industrial and manufacturing phase.

Information is recognized as the most important commodity for development. Information literacy is important in health care, good governance and democracy, assiting in development of innovations that could generate income, and so forth.

Information technology assists in the fast distribution, storage and accessing of information. It will therefore have a direct impact on the development of economies and quality of life. Can Africa afford not to enter the information technology phase? How does Africa do so? Can it miss the industrial or manufacturing phase and leap directly into the information technology phase? What role would science and technology education play under such circumstances?

# Appendix 1

## List of discussants

| Name | Address | Tel/Fax |
|---|---|---|
| Acquaye-Brown, Henry (Dr) | University of Cape Coast<br>Cape Coast<br>Ghana | 233-04-232480 (w)<br>233-04-233862 (h) |
| Anamuah-Mensuah, Jophus (Prof) | University of Cape Coast<br>Cape Coast<br>Ghana | 233-04-232480 (w)<br>233-04-232449 (h) |
| Cobern, Bill (Prof) | College of Education<br>Arizona State University<br>PO Box 37100<br>Phoenix<br>Arizona | 602-54-36300 (w)<br>602-54-36350 (fax) |
| Cole, Magnus (Dr) | Nyala University College<br>Freetown<br>Sierra Leone | |
| Dyasi, Hubert (Dr) | City College of New York<br>Convent Avenue<br>138th Street<br>New York<br>NY 10031 | 212-6-508436 (w)<br>212-6-506970 (fax) |
| Gray, Brian (Dr) | University of Western Cape<br>Private Bag X17<br>Bellville 7535 | 021-9592649 (w)<br>021-9592647 (fax) |
| Hodzi, Richard (Dr) | Science Education<br>Program Specialist<br>UNESCO<br>Subregional Office<br>PO Box HG 435<br>Highlands Harare<br>Zimbabwe | 733497 (w)<br>733021 (fax)<br>(e-mail): uhar@unesco.org |

| | | |
|---|---|---|
| K'Opiyo, Francis (Dr) | Head of Department<br>c/o Rockefeller Foundation<br>Nairobi<br>Kenya | 2542-228061 (w) |
| Kahn, Michael (Dr) | Policy Analyst Centre for<br>Education Policy Development<br>PO Box 31892<br>Braamfontein<br>SA | 011-4036131 (w)<br>011-3393455 (h)<br>(e-mail):<br>wb14@act2.pwv.gov.za |
| Katama, Agnes (Ms) | Rockefeller Foundation<br>African Forum for Children's<br>Literacy<br>PO Box 4753<br>Nairobi<br>Kenya | 254-2-88061 (w) |
| Kyle, Bill (Dr) | Director<br>School of Maths<br>& Science<br>Department of Curric Instruct<br>Purdue University<br>West Lafayette<br>7907-1442<br>USA | 494-7935 (w)<br>496-1622 (fax) |
| Lewin, Keith (Prof) | EDB Institute of Education<br>University<br>of Sussex<br>BN19RGUK | 01273-606755 (w)<br>01273-678568 (fax)<br>(e-mail):<br>k.m.lewin@sussex.ac.uk |
| Magi, Thembi (Dr) | Head of Department<br>University<br>of Zululand<br>Private Bag X1001<br>Kwadlangezwa<br>SA | 0351-93911 (w)<br>0351-93149 (fax)<br>(email):<br>nvmagi@pan.uzulu.ac.za |
| Mhlongo, Nathi (Mr) | Coordinator<br>Primary Science Project<br>PO Box 51236<br>Musgrave<br>Durban<br>4062 | 202-8090 (w)<br>202-8095 (fax) |

| | | |
|---|---|---|
| Mulemwa, Jane (Dr) | Makere University<br>PO Box 7062<br>Kampala<br>Uganda | 256-532924 (w)<br>256-41-542542 (h)<br>256-41-530756 (fax) |
| Ogunnyini, MB (Prof) | Head of Department<br>University of Western Cape<br>Private Bag X17<br>Bellville<br>SA | 021-959 2525 (w)<br>021-951 2602 (fax) |
| Putsoa, Bongile (Dr) | University of Zululand<br>Kwaluseni Campus<br>Private Bag 4<br>Kwaluseni<br>Swaziland | 268-84011/85108 (w)<br>268-85276 (fax) |
| Seephe, Sipho (Prof) | Venda University<br>Private Bag X5050<br>Thohoyandou<br>Venda<br>SA | 0159-210 71 ext 2235 (w)<br>0159-2204 5 |
| Shankerdass,<br>Sharad (Dr) | Box 40043<br>Nairobi<br>Kenya | 254-2-581030 (w)<br>254-2-521681 (h) |
| Zesaguli,<br>Josephine (Dr) | Bindura University College for<br>Maths and Science Education<br>University of Zimbabwe<br>Private Bag 1020<br>Zimbabwe | 263-71-7531/3 (w)<br>263-71-7534 (fax) |

# Appendix 2

## List of participants

| | | |
|---|---|---|
| Prof Svein Sjoberg | University of Oslo | Oslo, Norway |
| Dr Mohammed Bilal | University of Dar-es-Salaam | Dar-es-Salaam, Tanzania |
| Dr Eddah Gachukia | Forum for African Women Educationalists | Nairobi, Kenya |
| Lorna Muraga | Forum for African Women Educationalists | Nairobi, Kenya |
| Dr Khotso Mokhele | Science & Technology Ethos | Pretoria, South Africa |
| Dr David Court | Rockefeller Foundation | Nairobi, Kenya |
| HF Gonthi | Malawi Institute of Education | Domasi, Malawi |
| Dr Pamela Greene | Sierra Leone Home Economics Association | Freetown, Sierra Leone |
| AD Sozi | The Aga Khan Primary School | Kampala, Uganda |
| Dominic DB Enjiku | Institute for Teacher Research | Kampala, Uganda |
| Hussein S Khatib | Science Camp Project Coordinator | Zanzibar, Tanzania |
| Suleman Rashid Seif | Kijangwani | Zanzibar, Tanzania |
| Pius B Ngeze | Kagera Writers' & Publishers' Cooperative Society | Dar-es-salaam, Tanzania |
| Sebtuu M Nassor | Principal Secretary, Dept of Curriculum Studies | Zanzibar, Tanzania |
| Nellie Mbano | Chancellor College | Zomba, Malawi |
| Hau Simion | Science Teachers' Association of Malawi | Zomba, Malawi |
| Carl Bruessow | The Wildlife Society of Malawi | Zomba, Malawi |

| | | |
|---|---|---|
| Dr MN Chilambo | Chancellor College | Zomba, Malawi |
| Esther Nicholson | PAMET | Blantyre, Malawi |
| Sifiso Ndimande | University of Durban-Westville | Durban, South Africa |
| Dr Stella Y Erinosho | Ogun State University | Ijebu-Ode, Nigeria |
| Prof James Okuta | Ahmade Bello University | Zaira, Nigeria |
| Prof Tolulope Wale Yoloye | University of Ibadan | Ibadan, Nigeria |
| Davinder Lamba | Mazingira Institute | Nairobi, Kenya |
| Leonard Mwashita | Zimbabwe Teachers' Association | Harare, Zimbabwe |
| Steve Murray | *Action Magazine* | Harare, Zimbabwe |
| Mamotena Mpeta | National University of Lesotho | Maseru, Lesotho |
| Margaret Keogh | Science Curriculum Initiative of SA | Durban, South Africa |
| Dr D Botes | FEDU Foundation | Gabarone, Botswana |
| Prof Adjepong | University of Cape Coast | Cape Coast, Ghana |
| Dr MA Isahakia | National Museum of Kenya | Nairobi, Kenya |
| Felicity Leburu | Ministry of Education | Gabarone, Botswana |
| James Chima | Wildlife Society of Malawi | Blantyre, Malawi |
| Dr Ray Charakupa | University of Botswana | Gabarone, Botswana |
| Dr Ole Popov | National Institute for Educational Development | Maputo, Mozambique |
| Dr Karen L Worth | Urban Elementary Science Project | Pasadena, USA |
| Prof Philip Morrison | Cambridge University | Massachusetts, USA |
| Prof Eleanor Duckworth | Cambridge University | Massachusetts, USA |

| | | |
|---|---|---|
| Prof Jerry Pines | California Institute of Technology | California, USA |
| Prof Robert V Lange | Brandeis University | Massachusetts, USA |
| Dr Gary Knamiller | University of Leeds | Leeds, UK |
| Peter Towse | University of Leeds | Leeds, UK |
| Prof Terry Russell | University of Liverpool | Liverpool, UK |
| Dr Fred Lubben | University of York | York, UK |
| Dr Bob Lange | Brandeis University | Gabarone, Botswana |
| Margaret M Komba | Ministry of Science, Technology & Higher Education | Mbeya, Tanzania |
| Ts'epo Ntho | Handspring Trust for Puppetry in Education | Johannesburg, South Africa |
| Abdallah Omer El Farra | Sana'a University | Yemen, Arab Republic |
| Dr James Toale | Foundation for Research & Development | Pretoria, South Africa |
| Dr Prince Nevathulo | Foundation for Research & Development | Pretoria, South Africa |
| Justin Dillon | King's College | Cambridge, UK |
| Magnus Cole | Nyala University College | Freetown, Sierra Leone |
| Willy Mwakapenda | Chancellor College | Zomba, Malawi |
| Ellen Mulaga | Chancellor College | Zomba, Malawi |
| Francis Maria Janurio | National Institute for Educational Development | Maputo, Mozambique |
| Jack Holbook | ICASE | Limassol, Cyprus |
| Mark Poston | Chancellor College | Zomba, Malawi |
| John Rogan | Western Montana College, Montana University | Dillon, USA |

| | | |
|---|---|---|
| Sharad Shankerdass | | Nairobi, Kenya |
| Bettina Walther-Njoroge | | Nairobi, Kenya |
| Prof Peter Okebula | Lagos State University | Lagos, Nigeria |
| Winnie Byanyima | | Nairobi, Kenya |
| Hal Dorf | North Michigan University | Michigan, USA |
| Anita Westerstrom | | Sweden |
| Mr Devathasen | Gauteng Department of Education | Marshalltown, South Africa |
| Dr PA Motsoaledi | Northern Transvaal Education Department | Pietersburg, South Africa |
| Saphiwe Belot | Orange Free State Educational Department | Bloemfontein, South Africa |
| Dr NC Manganyi | Director-General, Education Department | Pretoria, South Africa |
| Zizwe Balindela | Minister of Education, Eastern Cape | Bisho, South Africa |
| David Mabuza | Minister of Education, Mpumalanga | Nelspruit, South Africa |
| Tina Joemat | Minister of Education, Northern Cape | Kimberley, South Africa |
| Mamoekoena Gaoreteleve | Minister of Education, North West Province | Mmabatho, South Africa |
| Dr VT Zulu | Minister of Education, KwaZulu-Natal | Ulundi, South Africa |
| Dr H du Toit | Gauteng Department of Education | Braamfontein, South Africa |
| Prof Ahmed Bawa | Natal University, Pietermaritzburg | Pietermaritzburg, South Africa |
| Dr Triegaardt | Executive Director, NGO | Johannesburg, South Africa |

| | | |
|---|---|---|
| Lebs Mphahlele | Interim National Science Teachers' Association | Johannesburg, South Africa |
| Rob O' Donogue | Natal Parks Board | Pietermaritzburg, South Africa |
| Dr Nkonzo Mtshali | FULCRUM | Durban, South Africa |
| Peter Glover | Primary Science Project | Durban, South Africa |
| Diane Raubenheimer | Primary Science Project | Durban, South Africa |
| Diane Grayson | SAARMSE | Pietermaritzburg, South Africa |
| Tema Botlhale | PROTEC | Johannesburg, South Africa |
| Prof RK Appiah | Engineering Faculty, University of Durban-Westville | Durban, South Africa |
| Faroon Goolam | SEDP, University of Durban-Westville | Durban, South Africa |
| David Brookes | SEDP, University of Durban-Westville | Durban, South Africa |
| Allan Pillay | Science Education Division, University of Durban-Westville | Durban, South Africa |
| Tholang Z Maqutu | Science Education Division, University of Durban-Westville | Durban, South Africa |
| Prof Jonathan Jansen | Education Faculty, University of Durban-Westville | Durban, South Africa |